Plasma Theory, Wave Kinetics, Nonlinear Dynamics: KaufmanFest 2007

Journal of Physics: Conference Series Volume 169

Berkeley, California, USA
5-7 October 2007

ISBN: 978-1-61738-314-4

Printed from e-media with permission by:

Curran Associates, Inc.
57 Morehouse Lane
Red Hook, NY 12571
www.proceedings.com

Some format issues inherent in the e-media version may also appear in this print version.

Printed by Curran Associates, Inc. (2010)

Institute of Physics
Dirac House, Temple Back
Bristol BS1 6BE UK

Phone: 44 1 17 929 7481
Fax: 44 1 17 920 0979

techtracking@iop.org

TABLE OF CONTENTS

Journal of Physics: Conference Series **169** (2009) 012001 doi:10.1088/1742-6596/169/1/012001

KaufmanFest 2007: Plasma Theory, Wave Kinetics, and Nonlinear Dynamics

A. J. Brizard[1], **E. R. Tracy**[2]

[1] Saint Michael's College, Colchester, VT 05439, USA

[2] Department of Physics, College of William and Mary, Williamsburg, VA 23187-8795, USA

Abstract. A symposium in honor of Professor Allan N. Kaufman's 80th year was held at the University of California at Berkeley on October 5-7, 2007. The meeting celebrated Allan's contributions to plasma physics as well as his friendship and guidance. The present paper very briefly summarizes the talks presented.

Figure 1. Allan N. Kaufman at work in Building 50 at Lawrence Berkeley National Lab in the late 1970's shortly after the K-χ PRL paper (with John Cary) was accepted. (Photo courtesy of John Cary)

Journal of Physics: Conference Series **169** (2009) 012001 doi:10.1088/1742-6596/169/1/012001

At least once in one's professional career, it is a useful exercise to stop and look back at what one has achieved. We look at our actions in terms of our own achievements and the people we have interacted with and influenced. While the former actions might be a worthy source of pride in one's skills, it is often the latter actions that have the most far-reaching impact.

As a teacher, mentor, and plasma physicist over the past 50 years, Allan Kaufman's influence in theoretical plasma physics cannot be exaggerated. As a teacher at UC Berkeley, his lectures in plasma theory were masterfully crafted by his desire to uncover the most fundamental aspects of plasma physics. As a mentor, many young plasma theorists can attest to what Allan's interest and support meant to them early in their careers. As a plasma theorist, Allan's constant search for elegance and simplicity in his work have become a signature that is, unfortunately, not often replicated.

As a celebration of Allan's contributions to plasma physics, as well as his friendship and guidance, a symposium in honor of Professor Allan N. Kaufman's 80th year was held at the University of California at Berkeley on October 5-7, 2007.

1. KaufmanFest 2007

A summary of the talks presented at the symposium is given below, drawn largely from the abstracts provided by the authors. The talk titles with a '*' have associated papers that appear in these proceedings.

*1.1. A. J. Brizard: Variational principles in plasma physics**

Alain Brizard (Saint Michael's College) began the symposium with a review talk about variational principles in plasma physics. The talk was both historical and conceptual, emphasizing the seminal role played by Allan Kaufman in either developing these ideas, or bringing them into plasma physics from other fields such as mathematics.

*1.2. K. Sonnad: Analyzing charged particle beams using Lie transform perturbation theory**

The plenary talk was followed by one by Kiran Sonnad (Center for Beam Physics, LBNL) concerning a recent application of Lie transform perturbation theory to analyze the dynamics of single charged particle motion in beams as well as for beams with space charge effects. In the first case, the analysis showed the existence of a condition that leads to improved integrability for a nonlinear focusing system. Numerical tracking of the particles showed that this condition provided an improved dynamic aperture. This analysis was then extended to study beams with space charge effects satisfying equations governed by a Vlasov-Poisson system. The latter work was used in calculating equilibrium phase space distribution functions.

1.3. B. A. Shadwick: Structure Preserving Numerical Methods for Differential Equations

Brad Shadwick (University of Nebraska) then spoke about new methods for the numerical study of differential equations. In physics one regularly encounters systems of differential equations that have a non-trivial dynamic or kinematic structure: the system may possess algebraic or differential invariants; the time advance transformation may belong to some symmetry group; the dynamics may take place on a manifold with non-trivial geometry; etc. Rarely will this structure not have a physical significance. Often this structure is responsible for important qualitative features of the system (and indeed may have purposely been built into the system). Generic numerical methods for solving initial value problems typically do not preserve any structure possessed by the system. By not preserving this structure, a numerical solution is generated which is allowed to exhibit behaviour denied to the physical system. The quintessential example of the importance of structure preserving methods is the success of symplectic integrators in accelerator physics and astronomy. A common theme of the talk was the computational efficiency gained by algorithms specifically designed to explicitly honor a system's structure.

Journal of Physics: Conference Series **169** (2009) 012001 doi:10.1088/1742-6596/169/1/012001

1.4. G. M. Webb: Variational Approach to Nonlinear, Oblique, Traveling Waves in Two-Fluid, Charge-Neutral, Electron-Proton Plasmas
Gary Webb (Institute of Geophysics and Planetary Physics, UC Riverside) then gave a talk about variational approaches to nonlinear waves in a two-fluid theory. Dual variational principles for nonlinear, oblique, traveling waves in charge-neutral, electron-proton plasmas were discussed. The equations are written as a dual spatial Hamiltonian system. The system is integrable in the case where the total transverse momentum integrals are zero in the de Hoffman-Teller frame of MHD shock theory.

1.5. C. Grebogi: Chaotic dynamics and modeling
The second plenary talk was given by Celso Grebogi (King's College, University of Aberdeen). Scientists attempt to understand physical phenomena by constructing models. A model serves as a link between scientists and nature, and one basic goal is to develop models whose solutions accurately reflect the nature of the physical process. But a dynamical model uses simplifying assumptions and approximations. The question of whether a model accurately reflects nature is one constantly faced by scientists. Celso argues that there exist levels of mathematical difficulty, brought from the theory of dynamical systems, which can limit our ability to represent chaotic processes in nature using deterministic models.

1.6. A. Lichtenberg: Nonlinear dynamics of oscillator chains
Allen Lichtenberg (Electrical Engineering and Computer Science Department, UC Berkeley) then spoke. The Fermi-Pasta-Ulam (FPU) nonlinear oscillator chain has proved to be a seminal system for investigating problems in nonlinear dynamics. First proposed as a nonlinear system to elucidate the foundations of statistical mechanics, the initial lack of confirmation of the researchers expectations eventually led to a number of profound insights into the behavior of high-dimensional nonlinear systems. Some of these insights, both old and new, were described.

*1.7. J. A. Krommes: Nonlinear theoretical tools for fusion-related microturbulence: Historical evolution, and recent applications to zonal-flow dynamics**
Fusion poses an extremely challenging, practically complex problem that does not yield readily to simple paradigms. Nevertheless, many of the theoretical tools and conceptual advances emphasized by Prof. Kaufman and others at the symposium have motivated and/or found application to the development of fusion-related plasma microturbulence theory. A brief historical commentary was given by John Krommes (PPPL, Princeton University) on some aspects of that specialty, with emphasis on the role (and limitations) of Hamiltonian/symplectic approaches, variational methods, oscillation-center theory, and nonlinear dynamics. The modern problem of zonal-flow dynamics illustrates a confluence of several of those methods, including (i) the use of Hamiltonian formalism to determine the appropriate (Casimir) invariant and wave-kinetic formalism for systems of interacting zonal flows and drift waves; and (ii) the application of nonlinear dynamics methods, especially center manifold theory, to the problem of the transition to plasma turbulence in the face of self-generated zonal flows. Finally, some unsolved problems of current interest (including intermittent chaotic statistics and the generation of coherent structures from turbulence) were mentioned, and an appeal was made for some new tools to cope with these interesting problems in nonlinear plasma physics.

1.8. C. S. Liu: Acoustic and gravity waves in dusty plasma
Chuan Liu (Physics Department, University of Maryland) gave a talk about acoustic and gravity waves in dusty plasma. A review was given of both theory and experiment concerning these topics with a particular emphasis given to wave-particle resonance.

1.9. G. J. Morales: Alfvénic phenomena triggered by resonant absorption of an O-mode pulse
George Morales (Physics and Astronomy Department, UCLA) then spoke about a simulation and modeling study of the nonlinear interaction of an electromagnetic pulse, in the O-mode polarization, with a magnetized plasma having a cross-field density gradient. For small amplitudes the pulse propagates up to the cutoff layer where an Airy pattern develops. Beyond a certain power level the ponderomotive force produced by the standing electromagnetic fields carves density cavities. The excess density piled-up on the side of the cavities causes secondary, field-aligned plasma resonances to arise. Strong electron acceleration occurs due to the short-scale of the secondary resonant fields. The fast electrons exiting the new resonant layers induce a return current system in the background plasma. This generates a packet of shear Alfvén waves of small transverse scale and increasing frequency. The results provide insight into microscopic processes associated with a recent laboratory investigation in which large amplitude Alfvén waves have been generated upon application of high-power microwaves.

*1.10. R. L. Dewar: The evolution of a meme: the quest for an oscillation-center variational principle**
Bob Dewar (Department of Theoretical Physics and Plasma Research Laboratory, Australian National University) gave a historical talk on the development of a variational principle for oscillation center formulations. The Lie transform approach was originally introduced in the hope of finding a canonical transformation, continuously connected to the identity, that would reduce nonresonant wave-particle interactions to a smoothly varying ponderomotive Hamiltonian quadratic in the wave amplitude, leaving only an irreducible strong interaction between waves and particles due to wave-particle resonances. A variational principle was later proposed in the hope of making such a separation between "resonant" and "nonresonant" unique. Although never very successful in the context of weak turbulence theory, this meme moved into KAM theory and now forms part of the memetic material in a current body of work on 3D MHD equilibrium theory.

1.11. M. Porkolab: Mapping of RF Waves in Tokamak Plasmas by Phase Contrast Imaging
Miklos Porkolab (Department of Physics and Plasma Science and Fusion Center, MIT) discussed recent progress in measuring RF waves in the ion cyclotron range of frequencies in the Alcator C-Mod tokamak by means of the Phase Contrast Imaging (PCI) diagnostic. This diagnostic allows one to map the radial extent of mode converted plasma waves, including their dispersion and direction of propagation. By such techniques the MIT researchers have discovered that at the ion-ion hybrid layer in a two ion species plasma, mode conversion into kinetic ion cyclotron waves (ICW) may dominate conversion into ion Bernstein waves (IBW). This is a consequence of strong magnetic shear off the mid-plane in tokamak plasmas. The results are in good agreement with earlier theoretical calculatiions and more recent full wave code predictions. In principle, using a calibrated PCI diagnostic, power flow among the different waves inside the tokamak plasma could be measured, thus providing a complete test of theoretical models.

1.12. B. I. Cohen: Hamiltonian Methods, Action-Angle Variables, Perturbed Trajectory Calculation, Nonlinear Dynamics and Quasilinear Theory Under the Influence of Allan Kaufman
Bruce Cohen (LLNL) provided a retrospective review of the calculation published in G. R. Smith and B. I. Cohen, Phys. Fluids 26, 238 (1983). This calculation used Hamiltonian methods, action-angle variables, and perturbed trajectory methods to derive a quasilinear theory for ion-cyclotron fluctuations in mirror machines. The dominant decorrelation mechanism turned out to be ion-bounce resonance overlap determined by a Chirikov analysis of the nonlinear dynamics. This calculation (performed by two of Allan Kaufman's Ph.D. students in the 1970s)

Journal of Physics: Conference Series **169** (2009) 012001 doi:10.1088/1742-6596/169/1/012001

was profoundly influenced by Allan's research and exemplified his research interests and the methods that he developed.

1.13. Bedros Afeyan: Kinetic Electrostatic Electron Nonlinear (KEEN) Waves in Vlasov Plasmas

Bedros Afeyan (Polymath Research) elucidated the properties of nonlinear non-stationary self-organized asymptotic (NNSA) states in plasmas which have no fluid nor a linear, small amplitude limit. These states are truly nonlinear constructs in phase space involving wave-particle interactions that are not strictly local in space and time. As particles give energy back to waves which trap them, and are, in turn, accelerated by those waves, the multimode (multiple harmonic) field structures that are created self-consistently, need not either trap once and for all or leave a particle untrapped forever. Instead, trapping, untrapping and retrapping oscillations occur which allow a given particle to interact intermittently with the self-consistent field (of all the particles) at different locations and with different outcomes. This leads to a partitioning of phase space into a few significant regions with different breathing trapping fractions, each of which has its own unique temporal signature. The collective action of these phase-space partitions is to maintain a long lived (undamped) self-consistent, non-stationary state in the plasma that is robust. KEEN waves are an example of NNSA structures that continuously span the frequency range somewhat below the electron plasma wave (EPW) frequency to around a tenth that frequency for all $k\,\lambda_D$ values up to order one, and for reasonable (finite duration) drive amplitudes, where conventional wisdom would have said no sustainable waves of this nature should have been possible. The multimode structure of NNSA states is a key feature as is their non-stationarity. These two features allow the interaction between KEEN waves, for example, and conventional weak amplitude and delicate waves such as EPWs to be very strong and nonlocal in phase space. Under the right conditions KEEN waves can siphon off energy from EPWs and thus turn off undesirable instabilities such as Raman and Two Plasmon Decay that rely on EPWs and which are of considerable worry in Inertial Fusion Energy applications and High Energy Density Laboratory Plasma Physics. The pioneering work of Kaufman and his students starting in the early 1970's was instrumental in the creation of this field of research where the kinetic properties of plasmas play such a central role, where nonlinear dynamics and statistical mechanics meet and get stirred up by coherent laser energy. This work was done in collaboration with Mathieu Charbonneau-Leford, Marine Mardirian, Vlad Savcheko, Kirk Won (*Polymath Research, Pleasanton, CA*) and Magdi Shoucri (*IREQ, Quebec, Canada*) and was supported by the DOE NNSA SSAA Grants program.

1.14. A. Bhatacharjee: Landau damping in weakly collisional plasmas and BGK waves

Amitava Bhatacharjee (University of New Hampshire) then spoke about progress he and co-workers have made to the understanding of the effects of collisions on Case-van Kampen (CvK) modes. They have shown that the CvK method for constructing solutions of the initial value problem is significantly modified and the CvK continuum is destroyed by the perturbation. This is because the collision term is a singular perturbation. Amitava spoke about more recent results where these ideas have now been extended to included nonlinear effects and BGK waves.

1.15. R. R. Lindberg: Nonlinear Langmuir Waves in Slowly-Varying Thermal Plasmas

Ryan Lindberg (Physics Department, UC Berkeley) gave a talk about the longitudinal dynamics of a resonantly driven Langmuir wave. The dynamics are analyzed in the limit that the growth of the electrostatic wave is slow compared to the bounce frequency. Using simple physical arguments, the nonlinear distribution function is shown to be nearly invariant in the canonical particle action, provided both a spatially-uniform term and higher-order spatial harmonics are included along with the fundamental in the longitudinal electric field.

1.16. C. B. Schroeder: Warm Wavebreaking of Nonlinear Laser-Driven Electron Plasma Waves
This talk was followed by one from Carl Schroeder (L'OASIS Group, LBNL). Short-pulse (sub-ps), intense ($> 10^{18}\ W/cm^2$) laser-plasma interactions are in a regime where the plasma electrons undergo relativistic motion while the plasma temperature (momentum spread) remains small. The nonequilibrium plasma is typically created by the laser through photoionization of a gas, and the laser-plasma interaction occurs on a timescale short compared to the ion motion and the collision frequency. The nonlinear Langmuir waves excited by intense lasers propagating in underdense plasma are analyzed using a warm, relativistic fluid model of a nonequilibrium, collisionless plasma. Properties of the nonlinear electron plasma waves, such as the plasma temperature evolution and nonlinear wavelength, are examined, and the maximum amplitude of the nonlinear oscillation (wavebreaking field) is derived. Electron plasma waves with relativistic phase velocities driven by intense short-pulse lasers can accelerate charged particles, and the maximum plasma wave amplitude sets a limit to the achievable accelerating gradient.

*1.17. E. R. Tracy: Ray-based methods and linear mode conversion**
Gene Tracy (William and Mary) gave a plenary review talk about ray-based methods and linear mode conversion. A survey of a few of the many contributions that Allan Kaufman has made to the study of waves in plasmas and fluids was given, with an emphasis upon the use of ray-based (WKB) methods. The talk was largely historical and conceptual rather than technical. Themes included the use of geometrical pictures as a guide for asymptotic approximations, the use of symmetries and action principles as a means to derive conservation laws, and examples of the fruitful use of modern mathematical ideas to solve physics problems. Applications that were briefly touched upon include RF heating in tokamaks, equatorial waves in the Atlantic, neutrino oscillations, and magnetohelioseismology.

1.18. A. Jaun: Ray tracing for ICRF, including mode conversion and caustics
Andre Jaun (Royal Institute of Technology, Stockholm) spoke about work he has done with Allan to develop practical ray-tracing codes that are based upon many of the ideas Allan has developed. Ray optics is used to model the propagation of short electromagnetic plasma waves in toroidal geometry. The new RAYCON code evolves each ray independently in phase space, together with its amplitude, phase and focusing tensor to describe the transport of power along the ray. Particular emphasis was laid on caustics and mode conversion layers, where a linear phenomenon splits a single incoming ray into two. The complete mode conversion algorithm was described along with its use to study the two space dimensions that are relevant in a tokamak. Applications were shown using a cold plasma model to account for mode conversion at the ion-hybrid resonance in the Joint European Torus.

1.19. J. R. Cary: Nonlinear harmonic generation in EBW propagation
John Cary (Tech-X Corporation and Department of Physics and Center for Integrated Plasma Studies, University of Colorado at Boulder) then spoke about how second harmonic Electron Bernstein Waves (EBW) can be generated due to the self-interaction of the fundamental wave as it propagates into the plasma. With variation of plasma parameters, there is generically a point at which one can obtain matching of both the frequency and the wavenumber. Moreover, the energy transfer can be large, because the matching point can occur where the second harmonic has nearly zero group velocity, and so builds to large amplitude. Nonlinear ?f and full particle-in-cell (PIC) simulations show that the amount of power pulled out of the fundamental can be a significant fraction of power levels expected to be used in experiments. The resulting power can be deposited at the half sub-harmonic resonance of the plasma in a completely different spatial location.

1.20. A. E. Charman: A new variational principle for synchrotron radiation
Andrew Charman (Physics Department, UC Berkeley) gave a talk about a new variational principle for synchrotron radiation. Within the framework of a Hilbert space formalism, he derives a maximum-power variational principle applicable to classical spontaneous radiation from prescribed classical harmonic current sources. A simple derivation was provided for the case of three-dimensional fields propagating in vacuum, and specialization to the case of paraxial optics was discussed. The techniques have been developed to model undulator radiation from relativistic electron beams (for which an example involving high harmonic generation is reviewed), but are more broadly applicable to synchrotron or other radiation problems, and may generalize to certain structured media. This variational principle appears to be independent of other well-known variational principles in electrodynamics based on action or reaction.

1.21. E. Knobloch: Convectons
Edgar Knobloch (Physics Department, UC Berkeley) spoke about *convectons.* Recent simulations of binary fluid convection with a negative separation ratio reveal the presence of multiple numerically stable spatially localized steady states he calls "convectons". These states consist of a finite number of convection rolls embedded in a nonconvecting background and are present at supercritical Rayleigh numbers. Below a critical Rayleigh number the convectons are replaced by relaxation oscillations in which the steady state is gradually eroded until no rolls are present (the slow phase), whereupon a new convecton regrows from small amplitude (the fast phase) and the process repeats. Both He_3-He_4 mixtures and water-ethanol mixtures exhibit this remarkable behavior. Stability requires that the convectons are present in the regime where the conduction state is convectively unstable but absolutely stable. The multiplicity of stable convectons can be attributed to the presence of a 'pinning' region in parameter space, or equivalently to a process called homoclinic snaking. In the pinning region the fronts bounding the convecton are pinned to the underlying roll structure; outside it the fronts depin and allow the convecton to grow at the expense of the small amplitude state (large Rayleigh numbers) or shrink back to the small amplitude state (low Rayleigh numbers). The convectons may exist beyond the onset of absolute instability but the background state is then filled with small amplitude traveling waves. A theoretical understanding of these results was developed.

1.22. K. A. Mitchell: The Topology of Nested Tangles
Kevin Mitchell (School of Natural Sciences, UC Merced) then spoke about homoclinic and heteroclinic tangles. These are fundamental geometric objects that organize the phase space of chaotic systems, providing a framework for chaotic transport and escape processes. The topology of such tangles exhibits a striking richness in complexity and variability, which continues to be an active field of fundamental and applied research. Recently, Kevin and co-workers introduced a new topological technique, "homotopic lobe dynamics", that provides a systematic approach to defining symbolic dynamics for tangles of arbitrary complexity by inputting information about the short time structure of a general tangle, one can predict the resulting structure forced at later times. For area-preserving maps, the topology of a homoclinic tangle is typically complicated by the presence of stable island chains, which pull the tangle into their vicinity, thereby increasing the tangle complexity. He studies this process by considering the heteroclinic tangles attached to the hyperbolic periodic orbits situated between the stable islands. This results in a system of homoclinic and heteroclinic tangles nested within each other, to which he applies the techniques of homotopic lobe dynamics, producing an enhanced symbolic description of the map. The net result is to generate symbolic dynamics that serve as a topological approximation to the map that includes the influence of stable island chains. Such symbolic descriptions are useful in the study of fractal self-similarity in chaotic scattering and escape data, the computation of escape rates, the calculation of topological entropies, etc.

*1.23. P. J. Morrison: Thoughts on Brackets and Dissipation: Old and New**
Phil Morrison (UT Austin) gave a retrospective talk on the notion of the *dissipative bracket.* In 1981 Allan had the idea to formulate dissipation in terms of brackets with some algebraic properties, an extension of Phil's work on noncanonical Poisson brackets that he had begun with John Greene a couple of years before. Allan invited Phil to join him and this resulted in a paper on quasilinear theory. By 1984 they had gone our separate ways on this topic. Later, two papers appeared by other authors that introduced the double bracket for describing a kind of dissipation. In this talk Phil reviewed some of the above, talked about some old ideas he had in the mid-1980's that were never published, and described what happened when he started thinking about them again while preparing this talk.

1.24. S. Omohundro: Self-Improving Artificial Intelligence and the Future of the Universe: Where Physics Meets Computation
The last talk of the symposium was by Steve Ohmohundro (Self-Aware Systems). His topic concerned self-improving artificial intelligence and the connection between physics and computation. Can we predict the behavior of systems that modify themselves? Can we design them to embody our values even after many generations of self-improvement? This talk presented a framework for answering questions like these. It showed that self-improving systems converge on a specific cognitive architecture that arose out of von Neumann's foundational work on microeconomics. In these systems there is a universal principle which governs the organization of physical and computational resources at every level. These systems exhibit four natural drives: 1) efficiency, 2) self-preservation, 3) resource acquisition, and 4) creativity. Unbridled, these drives lead to both desirable and undesirable behaviors. The efficiency drive leads to algorithm optimization, data compression, atomically precise physical structures, reversible computation, adiabatic physical action, the virtualization of the physical, and governs a system's choice of memories, theorems, language, and logic. The self-preservation drive leads to defensive strategies such as "energy encryption" for hiding resources and promotes replication and game theoretic modeling. The resource acquisition drive leads to a variety of competitive behaviors and promotes rapid physical expansion and imperialism. The creativity drive leads to the development of new concepts, algorithms, theorems, devices, and processes. The best of these traits could usher in a new era of peace and prosperity; the worst are characteristic of human psychopaths and could bring widespread destruction. How can we ensure that this technology acts in alignment with our highest values? We have leverage both in designing the system's initial values and in creating the social context within which they operate. But we must have great clarity in imagining the future we want to create. We need not just a logical understanding of the technology but a deep introspection into what we cherish most. With both logic and inspiration we can work toward building a technology that empowers the human spirit rather than diminishing it.

2. Summary
As this summary shows, it was a very stimulating two days of talks. The range and depth of the topics reflects the impact that Allan has had either directly or indirectly on the conference participants, many of whom have worked directly with Allan, or admired his work from afar.

The conference organizers would like to take this opportunity to thank all the speakers, and the participants who presented posters, for making this a memorable meeting.

Plasma Theory, Wave Kinetics, Nonlinear Dynamics: KaufmanFest 2007 IOP Publishing
Journal of Physics: Conference Series **169** (2009) 012002 doi:10.1088/1742-6596/169/1/012002

A half-century in plasma physics

Allan N. Kaufman
Department of Physics and Lawrence Berkeley National Laboratory, University of California, Berkeley

Abstract. This memoir is an autobiography of my life as a plasma theorist, highlighting the many individuals who contributed to my intellectual development.

My life in plasma physics began in 1955, when Dick Post gave a lecture at the Livermore Radiation Lab (now LLNL) on the (classified) Sherwood program for controlled thermonuclear energy, and on the mirror-machine research which he directed. (Sherwood was named after Jim Tuck, one of its founders. Jim had been one of my experimental research supervisors at the University of Chicago.) The mirror machine was introduced by Edward Teller and Herb York, who also headed the nuclear weapons program at the newly founded Livermore lab. Edward, one of my professors at Chicago, had brought me to Livermore in 1953, to work on weapons.

Utterly fascinated by the scientific and applied aspects of Sherwood ("success in 20 years"), I asked my weapons-research supervisor Johnny Foster to arrange my transfer to Sherwood. And so this new life began. This was the ground floor of plasma physics, and I was hoping to contribute to its development.

In this memoir, I would like to recognize the many persons who interacted with me, and who provided the exciting stimulation for my scientific life. Let me begin with my pre-plasma life.

1. My parents

My parents were married in 1926, the year of the Schroedinger equation. I was born at the University of Chicago Lying-In Hospital in 1927, oblivious of the wonderful developments in physics at that time. Both my parents had emigrated to Chicago from Germany in 1910, for economic reasons.

My father Justin was born in Cochem in 1888, the one year (actually only 99 days) that Germany had a liberal Kaiser, Friedrich III, son-in-law of Queen Victoria [1]. Cochem is a charming town, among steep vineyards on the Mosel River in the Rhineland, and at the foot of a magnificent medieval castle, built in the 19th century. His father Arnold was a watchmaker; his mother Babette Rothschild (no relation) was born in Bavaria in 1858, and remembered hearing about Lincoln's assassination. In Cochem he received a sound high-school education from a Catholic institution, highlighted by 5 years of Latin. (He transmitted his love of Latin to me, with many quotations: "quamquam sunt subaqua ...") After working for a few years in Furth, in 1910 he came to Chicago, to work for a cousin in retail clothing. He joined the Continental Grain Co. as an accountant, and rose to the position of Treasurer, retiring in 1957. This international family-owned (the Fribourgs) concern, headquartered in Antwerp, bought grain from farmers, stored it in silos, and sold it to flour mills. In my teens, I worked as an office-boy in their Board of Trade office on LaSalle St, trying to understand the mysterious world of business.

Journal of Physics: Conference Series 169 (2009) 012002 doi:10.1088/1742-6596/169/1/012002

My mother Millie Low was born in 1890, in Friedberg, Hesse, a few miles north of Frankfurt. The Low clan was believed to be descended from Rabbi Loew of Prague, creator of the legendary golem. Her father Nathan was a cattle dealer; her mother Clara died when she was only 5 years old. Thus she was raised by an aunt, and had sad memories of her childhood. She was the youngest of five siblings; her eldest brother Edward was born in 1879 (the same year as Albert Einstein) and emigrated to Chicago at age 11, to work for his elder half-brother. My mother's formal education ended with elementary school, but she continued learning through her 90s. She came to Chicago in 1910 with her older sister Paula, and worked first as a governess, and then (after business college) as a stenographer. In the '30s, she taught English to German refugees at the local Y, and typed Braille textbooks for blind students.

My parents settled in Hyde Park [2], a largely German-Jewish area, one mile from the University, near many parks (Jackson Park, Washington Park, Madison Park, Farmers Green, East View Park, East End Park) and the beach of Lake Michigan, and with excellent public transportation (train, bus, street car). Jackson Park had been the location of the 1893 Columbian Exposition, and still had its magnificent Palace of Fine Arts (converted by Julius Rosenwald's philanthropy [3] to the Museum of Science and Industry, where I spent many happy hours) with its lagoon, the Japanese Garden on Wooded Island within the lagoon, the Iowa Building, and the beautiful gilded statue of Columbia.

2. My childhood

My earliest vivid memory was political, accompanying my mother to the polling booth in November 1932, to vote for FDR. He was inaugurated on March 4 ("We have nothing to fear but fear itself"), but what I remember is the previous day, my Dad telling me that it was 3/3/33. Intimations of interest in math.

By then Hitler had become Nazi dictator, and my parents decided to take their first trip back to Europe, to visit relatives in Frankfurt, Hamburg, and Vienna. At this impressionable age, I gloried in the trans-Atlantic voyage on SS Manhattan and Washington (later to be troop carriers) and the sights of Paris (Eiffel Tower), Rome (Castel Sant' Angelo, to be seen again in Tosca, and the Colosseum), Venice (pigeons in St Mark's Plaza, mosaics in the Cathedral, the Campanile, the Rialto Bridge), Vienna (the Ferris Wheel in the Prater), and Cochem. Nazi flags were everywhere in Germany; my parents told me to speak to them only in English.

I had been bilingual, speaking German to my grandmother, who had not mastered English. When someone told my mother that I spoke English with a German accent, she decided to stop communicating with me in German. So I lost my speaking knowledge of German, as well as of Italian, which I had learned from my sitter during our one week in Rome.

The 30s were stressful times for the Jewish community. My parents' resources were devoted to helping our relatives escape from Germany, and getting them established in Chicago. My mother was religious, and our family attended services regularly at KAM Temple, the oldest synagogue in the Midwest. It had two successive liberal intellectual rabbis, Joshua Loth Liebman (author of the best-seller "Peace of Mind" [4]) and Jacob J Weinstein (who was a speech-writer for Adlai Stevenson's campaign for president in 1952) [5]. At my mother's request, Rabbi Weinstein trained me for my Bar Mitzvah, the first at this reform synagogue.

Returning to 1933, I loved the new World's Fair, A Century of Progress (Chicago had become a city in 1833), especially the science exhibits. Then first grade began at Kenwood School (no public kindergarten in those days), with a succession of elderly, devoted, mainly Irish, spinster teachers, who instilled my love of learning. Of course, I inherited this primarily from my parents, nature and nurture. I devoured their collection of books: H G Wells' "An outline of History", van Loon, Shakespeare, Einstein.

Journal of Physics: Conference Series **169** (2009) 012002 doi:10.1088/1742-6596/169/1/012002

3. Hyde Park High School

Fast forward to 1940, when I entered Hyde Park High. The first year was at the Annex, where I had an inspiring Latin teacher Miss Kirby ("... cum libertate iustitiaque omnibus") and English teacher Miss Slocum (Julius Caesar: "you blocks, you stones, you worse than senseless things"; Herve Riel; Evangeline; Ivanhoe). In Algebra, I learned that pi equals 22/7 exactly.

Next year to the main building, where all my (scientifically-oriented) friends enrolled in ROTC to avoid Physical Education and to prepare for military service in the war, and learned math from the greatest teacher of them all, Beulah I ("for isosceles") Shoesmith. Our plane geometry was strictly from Euclid, and taught us mathematical logic and beauty. Then came log tables (not even slide rules), trig, advanced algebra, solid geometry, but no calculus. Our other superb teachers were Miss Rubetta Biggs in English (Macbeth: "when shall we three meet again", Pride and Prejudice, Silas Marner, Twelfth Night: "if music be the food of love"), Miss Baumgartner in Zoology, and Miss Lyons in US History. My chemistry teacher taught us about local politics, and so I learned the elements of chemistry on my own. In physics, we were strictly classical: Archimedes to Faraday, no Einstein or atomic structure. At our wonderful public library, I found Eddington, Jeans, and Heisenberg, but couldn't understand how AB could be unequal to BA.

4. University of Chicago

At age 17, I attempted to enlist in the Navy, as many friends had done, to take the Eddy Test and become radar technicians. But since I wore glasses, my induction was delayed for a year. Thus, in June 1944 (shortly after D-Day), I started college at the University of Chicago, as did almost all my classmates, living at home. For several weeks, I was simultaneously finishing my high-school classes, a challenging situation. But even more challenging was introductory Physics, taught by young Mario Iona (who years later won the Millikan Award for his teaching excellence). This class of course used calculus, which I was learning at the same time. In addition, the College required a set of masterly presented survey courses, instituted by Robert Maynard Hutchins & Mortimer Adler, and based on the Great Books [6]. For these, attendance and exams were optional; one's grade was determined solely by an eight-hour exam at the end of the school year. These classes included Humanities (Herodotus, Thucydides, Aristotle, Plato, Antigone, Hamlet, The Alchemist, School for Scandal, Addison & Steele, Tom Jones, Hume, Locke, The Brothers K, The Trial, The Spoils of Poynton, James Joyce), Social Sciences (Durkheim, Max Weber, Schumpeter), and Biological Sciences (botany, zoology, physiology; no DNA). They led to the PhB degree, my first of four at UC.

My walks between home and the University took me past Stagg Field, surrounded by barbed wire. Of course I had no idea what was happening at the Metallurgical Lab inside. Then in August 1945 the world was transformed, and a few days later, the war ended, with many lives saved. A few weeks before, I had turned 18, and awaited my draft notice. In October, I was inducted into the Navy, that service being chosen because of my one-year college education. After boot camp at Great Lakes, just north of Chicago, I was sent to BuPers (Bureau of Naval Personnel) in Arlington, VA, near the Pentagon. Because I had taken calculus, I was assigned to perform arithmetic for a captain who prepared budgetary estimates for Congress. With demobilization, I had served only 9 months, and then returned to college.

As my math courses (e.g., from Halmos & Kaplansky) became more advanced and abstract, I realized that math would not be my profession. Chemistry was ruled out by a course in quantitative analysis, which required my full time one quarter. Thus physics became my career choice by elimination. At Chicago, there was no separation between undergraduate and graduate courses; the physics curriculum was otherwise structured logically. So after my one-year introductory course, I found myself in competition with graduate students, mostly returned veterans, from other institutions. Taking effectively graduate courses in my second year, I felt

Journal of Physics: Conference Series 169 (2009) 012002 doi:10.1088/1742-6596/169/1/012002

woefully unprepared, so decided to take a year off, to catch up on my own by self-study.

The physics courses were taught by all the faculty, some of whom were superb teachers (Fermi, Wentzel, Zachariasen, Chandrasekhar, Platt, Goldberger, Sachs), while many others were hopeless as teachers, though outstanding in research. My first course from Enrico Fermi was mathematical physics, wisely offered by the physics department because of the abstractness of the math courses. My fellow students and I quickly realized that Fermi was unique in his teaching ability, so we enrolled or audited every course he taught. (One summer he taught three courses!) Just listening to him made one feel that one had complete understanding; his insights were amazing. He taught almost every course offered: solid state, thermodynamics, field theory, nuclear physics. Being exceptionally modest, he never used his own name for the many concepts named for him: thus fermions became Pauli particles.

My fellow students Art Rosenfeld, Jay Orear, & Bob Schluter asked Fermi for permission to publish their notes from his course on nuclear physics [7]. They asked me to write the chapter on cosmic rays. The diagrams I drew, based on Vallarta's theory, showed regions in the Earth's field where particles could be trapped. But no one foresaw the existence of the van Allen belt, discovered over a decade later.

While taking courses, I was interacting with the older cadre of graduate students, some of whom were being directed by Fermi and Teller. They were clearly a brilliant bunch, and an inspiration: Yang and Lee, Chew and Goldberger, Marshall Rosenbluth, Mel Gottlieb (my lab TA), Owen Chamberlain. In my own classes were (in addition to Rosenfeld, Orear, & Schluter) Nina Byers, John Firor, Bob Friauf, Burt Fried, Maurice Glicksman, Chas Goebel, Bob Keyes, Steve Moszkowski, Moe Scharff, Ron Sladek, Frank Solmitz, Ludwig Tannenwald, Sam Treiman, George Wetherill, and Gaurang Yodh.

As an experimental Research Assistant, in 1949 I was assigned to Jim Tuck, who with Lee Teng was designing an extractor for the proton beam of the newly built UC synchrocyclotron [8]. The theory involved an iterated mapping, a beautiful technique new to me. My role was to design the iron shims, mill them in the machine shop, and crawl between the pole pieces to measure the magnetic field distribution, using a Dicke flip-coil fluxmeter. When Jim left for Los Alamos (he was the leading expert on high-explosive lenses, and had been the major British contributor to the Manhattan Project), John Marshall became my supervisor. John too later became prominent in Sherwood.

After passing the 32-hour departmental comprehensive physics exam (four 8-hour days) in 1951, I applied to Enrico Fermi for my PhD thesis research. He told me that he was sponsoring only experimental theses, and that I would first need to take an informal course on experimental technique from Dick Garwin, his postdoc. In spite of Dick's brilliance and patience, I was all thumbs, a danger to myself and others. So I was finally destined for theoretical research, and was accepted by Murph Goldberger (only 5 years my senior), who had recently completed his own thesis under Fermi's direction. Murph told me that you don't have to be a genius to be successful as a physicist. His first suggestion for a thesis topic was an extension to scattering of a variational principle developed by David Saxon (later to become president of the University of California system). When I reported my lack of progress to Murph, he said that he had no other thesis ideas, let's ask Gregor Wentzel. (I had taken Wentzel's wonderful course on quantum mechanics, in which he had modestly been unable to identify the W of the WKB method.) He told us that yes, he had a lowest-order result for meson-nucleon scattering in his strong-coupling theory of nuclear interaction (of 1940), and wanted someone to calculate the next-order contribution. This theory was the antithesis of the currently fashionable weak-coupling theory, based on Feynman diagrams. The fundamental technique was the use of noncanonical Poisson brackets, to be of great importance in my future plasma research. I studied Wentzel's original paper, found a serious error, corrected it, and proceeded with a lengthy and complicated calculation until I got stuck. Neither Murph nor Wentzel was able to resolve my difficulty, so Murph took me to

Journal of Physics: Conference Series **169** (2009) 012002 doi:10.1088/1742-6596/169/1/012002

young (2 years my junior) Murray Gell-Mann, who had just joined our faculty. Murray of course suggested the solution.

Toward the end of my thesis research, I faced the dreaded oral exam, covering current advanced topics. My committee consisted of Fermi, Wentzel, Murph, and Murray. (Murph wisely scheduled my exam when its experimental member, Herb Anderson, would be out of town.) To prepare, I asked my good friend Art Rosenfeld, who was working with Fermi, what were Fermi's current research ideas. Art provided me with his four latest preprints, none of which had yet been published. I studied them thoroughly, and thus was able to brilliantly answer Fermi's questions, to Murray's amazement. (I confessed to Murph afterwards.)

I was occupying Edward Teller's office and desk, as he was on leave to Los Alamos and setting up the Livermore lab. Murph recommended that I apply for a position in this new laboratory. Dave Judd and Joe Lepore came to Chicago to interview me, and soon I received an offer to join the weapons program, which I did in June 1953.

5. Weapons Research at Livermore

On arrival in Berkeley, I was greeted by Burt Fried and his wife Sally, who got me settled. Burt had recently completed his thesis with Wentzel, and was now a postdoc at the Rad Lab (UCRL), working with Bob Riddell. (Three years later, Bob would make a rare foray into plasma physics with Steve Gasiorowicz and Maurice Neuman, producing one of the first plasma kinetic equations.) I spent several weeks at UCRL, awaiting my security clearance. When it finally arrived, I was introduced to my new nuclear weapons colleagues, and was asked how I would design an atomic bomb. Of course I had no idea, and proceeded to learn the background theory in Courant & Friedrichs' wonderful textbook of 1948 [9]. My group leader was Johnny Foster, and my colleagues included Jim Wilson, Ed Chupp, Larry Germain, Chuck Godfrey, Jack Trulio, Wally Selove, Bryce DeWitt, Wally Birnbaum. I also interacted with Harold Brown, Hans Mark, Bob Jastrow, Sid Fernbach, Larry Wilets, Steve Brush, Mike May, Chuck Leith, Herb York, and of course Edward Teller. We felt that the Soviet threat was very real, and so we worked hard, often around the clock. Our computer was the Univac, utilizing the octal system and a punched tape input. (One motivation for its use was that it occupied the only air-conditioned room at the lab. For human brains, it was felt that Livermore summer temperatures would not affect productivity.)

One needed empirical datum was the high-pressure equation of state of the substance we used. The only person with the required instrument for reaching a megabar was Percy Bridgman of Harvard, who had received the Nobel in 1946 for this work. So I was appointed as the courier for a large sample of this radioactive pyrophoric substance, as carry-on baggage on a commercial flight to Boston. Prof Bridgman greeted me in his lab, and allowed me to participate in his measurements; he then flushed the sample down the drain. To determine the shock Hugoniot of this substance, I later designed an experiment using high explosives, which Wally Selove then carried out.

On a visit to Los Alamos, I was greeted by Jim Tuck, who showed me his Perhapsatron, utilizing the pinch effect, which again was totally new to me. (Somehow Amperian attraction of parallel currents had not been included in my E & M course at Chicago.)

While working at Livermore, I was living in Berkeley (with an 80-minute commute, before the introduction of freeways). So I could participate in evening activities on the campus, especially the weekly Physics Dept journal club. Here were presented reports on important new developments, and I was able to interact with Bill Nierenberg, Ernest Lawrence, & Luis Alvarez.

Journal of Physics: Conference Series **169** (2009) 012002 doi:10.1088/1742-6596/169/1/012002

6. Sherwood at Livermore

My transfer to the Sherwood program in 1955 represented a sudden reduction in pressure. (Our theory group leader, Ted Northrop, called our area Happy Valley.) My first task was to educate myself in this new field of high-temperature plasma physics. There were no textbooks yet published (Spitzer's excellent book came out in 1956 [10]). Because the program was secret (until 1958), there were few papers in the open literature. My main resource was a set of lectures given by Conrad Longmire in Los Alamos the year before.

There were good interactions among our group, which included Cliff Gardner (later known for his stability theorem and for his co-authorship of the inverse scattering method), Marvin Mittleman (whose wife Sondra arranged my introduction to my future wife Louise), Lewi Tonks (who had collaborated with Irving Langmuir in discovering Langmuir waves in 1928), Howard Greyber, Bill Newcomb, John Killeen (our mathematician), and later Nick Christofilos (inventor of the alternating-gradient or strong focusing cyclotron), Harold Furth (who was working with Stirling Colgate), and Bernie Lippmann.

My first publication for the Sherwood project was on collisional particle loss in the mirror machine, taking account of the electrostatic field caused by the excess of electron loss over ion loss. This was presented at a conference of 350 Sherwooders in Gatlingburg, in June 1956. Shortly thereafter, I spent a month at Los Alamos, having been invited by Murph and Ken Watson.

At the public library there, I found a book on how to select a spouse. This was a serious research topic, as I was emotionally committed to Louise Lazarus, whom I had met on Cinco de Mayo, thanks to my colleague Marv Mittleman and his wife Sondra. After a whirlwind courtship, I found in this book a checklist for marital compatibility, and confirmed my feeling that we were meant for each other. Louise accepted my invitation to travel to Mexico with the Mittlemans; in Mexico City, on Aug 19, at the romantic restaurant Rincon de Goya, I popped the question, and we were engaged. Our wedding took place in April 1957, at Temple Emanuel in San Francisco, by our rabbi Alvin Fine (author of "Life is a journey", cousin of Jacob Weinstein, and another inspiring liberal intellect). Louise quickly learned about life with a physicist, and has provided me with the moral support I needed throughout my career.

In early 1957, Stirling Colgate suggested that I examine the collisional transport of plasma mass and energy due to a temperature gradient perpendicular to a strong magnetic field. (The effect of density gradient was discussed in Spitzer's textbook.) The most important result of my calculation was that ion-ion collisions dominated thermal transport. After I had finished, Stirling confessed that he had suggested the same problem to Marshall. So I consulted Marshall, found that our results coincided, and wrote up our work for joint publication [11]. I followed this with a study of the effect of charge separation [12]. My next foray into transport theory was stimulated by Nick Christofilos' question as to the microscopic explanation for the formulas of magnetized-plasma viscosity. I found that shear flow implied elliptical gyro-orbits, and led to the desired explanation [13].

In the summer of 1957, and newly married, Louise and I spent some weeks in Los Alamos. I worked with Ken Watson and Chandra, formulating a fluid theory for collisionless plasma. Our major success was the discovery of the instabilities due to anisotropy of the pressure tensor [14].

In the early summer of 1958, we went to San Diego, where I worked at Convair (General Dynamics) with Ed Hones & George Stewart, and with Bob Vik, Bob Karplus, and Ken Watson, on the dynamics of an artificial belt of electrons in the Earth's magnetic field. The electrons would be injected by a nuclear explosion. This had been proposed by Nick Cristofilos, as a defense against Soviet ICBMs; his proposal was later tested in Operation Argus. While at San Diego, I was interviewed by Roger Revelle, who showed me the site he envisioned for the new UC campus, to open in 1960. But I opted to stay at the Livermore lab.

In September 1958, we proceeded to Geneva, for the Atoms for Peace Conference, at which the US, British, & Soviet fusion programs were first put on display and declassified. (Because the AEC had strangely decreed that only experimentalists could attend, I was given a crash course in operating one of the lab's plasma devices. Of course, at the conference itself, I was not allowed to touch it.) We met the leading Soviet theorists: Igor Tamm (who gave an after-dinner speech about his Menshevik days during the Revolution), Roald Sagdeev, Boris Kadomtsev, Evgeny Velikhov, Vitaly Shafranov.

On the day we returned to the states, we visited my father in the hospital, and he died a few hours later. My mother had protected us from the knowledge of his terminal illness.

In the summer of 1959, I was invited by Cecile Morette to lecture at the Les Houches summer school, near Chamonix & Mont Blanc, on the subject of plasma dissipative processes. My fellow lecturers included Leon Van Hove & Elliott Montroll. The students, half French and half foreign, were very bright; one of them was Donald Lynden-Bell.

7. Teaching at UC Berkeley

In the fall of 1959, the need arose for someone to teach the one-year graduate course in electromagnetic theory (Physics 210, E & M) on the Berkeley campus, as the professor assigned to it had left to take a sabbatical. With one week's notice, I was offered this position (a one-year contract, at one-third time, as Lecturer) by Carl Helmholz, the chairman. Bob Karplus, who had taught the course the previous year, gave me his lecture notes to use. And so my teaching career began.

Of course, I had to prepare the exams on my own. For the first mid-term exam, the class average was 55 out of 100; that seemed quite satisfactory. On the final exam, the class average was again 55, but this time out of 300. (A nice example of invariance.)

In using Panofsky & Philips as my textbook, I was dissatisfied by their treatment of dielectric energy. I expressed my discomfort to Ken Watson on a Friday; the following Monday, he came in with an extensive set of notes, laying out a statistical mechanical approach to this subject. The problem posed was the thermal equilibrium of an imperfect gas in an applied electric field. For an ideal gas (of identical molecules), there was a known result, the Clausius-Mossotti formula, relating the dielectric constant to the single-molecule polarizability. The challenge was to deal with both the short-range interaction between molecules and their long-range dipole-dipole interaction induced by the electric field. Of course one had to define the macroscopic field clearly, as opposed to the random microscopic field. Working with Ken on this problem was a great experience [15, 16]. Later we discovered a similar treatment in the Handbuch der Physik article by Wm Fuller Brown, as well as a sophisticated macrosopic theory by Landau, in terms of thermodynamic free energy. It was natural to then study the analogous magnetostatic problem, which I conducted with Ken's postdoc Toshio Soda [17]. Whereas the electrostatic problem allowed a classical treatment, a quantum formulation was required for the magnetic problem, since the classical magnetic susceptibility vanishes identically, by the Bohr-van Leeuwen theorem. We utilized the Darwin Lagrangian, which we had studied in earlier work [18] in connection with the definition of magnetic energy. Later, with Pete Rostler (Wulf Kunkel's student), we proposed the Darwin Lagrangian as an appropriate tool for plasma dynamics simulation [19].

Other papers resulted from perceived deficiencies in standard textbook treatments. On the relation between microscopic and macroscopic electric field, I found a subtle correction to Leon Rosenfeld's treatment, involving molecular quadrupole density [20]. For Maxwell equations in moving media, I was surprised that it had always been assumed (e.g., in Tolman's treatise) that the motion was spatially uniform and temporally steady, and that non-steady motion required a general-relativistic treatment. Accordingly, I developed a statistical approach based on local inertial frames, to define the macroscopic fields [21]. This was later generalized by DeGroot and Suttorp. Another question concerned the equation of motion of a dielectric medium, which Landau & Lifshitz had left open [22].

In the fall of 1961, Wulf Kunkel initiated and organized a series of technical lectures throughout the state on the extensive new developments in plasma physics, theoretical, experimental, and applied. He invited me to present the lecture on dissipative effects. He then edited the written versions of these lectures, in a book published in 1966 [23].

During the 1961-62 school year, the Livermore lab asked me to offer the same course 210 that I was teaching in Berkeley (MWF in Berkeley, TuTh in Livermore). My prize student was Gerry Schubert (now an eminent geophysicist), who commuted in from his full-time job at the Navy base in Vallejo.

When Lou Klauder requested that I serve as his PhD thesis advisor, I proposed that he extend our theory of linear polarization to higher order in the field, to exhibit electrostriction. His thesis [24] was completed in 1964. After Lou, my next three students all did plasma-physics theses: Shalom Fisher [25], Gary Pearson [26], and John C. Price [27].

By 1963, it was apparent to me that a graduate course in plasma theory was needed, to supplement the undergraduate course (Physics 142) that Wulf Kunkel was teaching. To prepare myself in modern plasma theory, I took a leave from the Livermore lab, to spend some time at the Princeton lab (PPPL), where many of the leading theorists worked: Ira Bernstein, John Dawson, Ed Frieman, John Greene, John Johnson, Martin Kruskal, Russell Kulsrud, Andy Lenard, and Carl Oberman. We rented Murph & Mildred Goldberger's beautiful house on Fitzrandolph Road. By then we had our two children: Joel, born in 1960; Janet, born in 1962. From our neighbor Evelyn Patterson, Louise acquired her classic "Recipes for elegant but simple dining".

Thus, in the fall of 1963, I offered the first semester of the graduate course Theoretical Plasma Physics 242A, limiting the material to unmagnetized plasma. Ted Northrop then presented the magnetized case, 242B, underlining B, in Spring 1964. This course was eventually expanded to four quarters or three semesters, and was offered every other year. While this physics course was analytic and based on kinetic theory, my colleagues in EE presented MHD theory (Allan Lichtenberg and Mike Lieberman) and numerical simulation (Ned Birdsall).

8. UCLA interlude

In early 1964, I received an offer to join the physics faculty at UCLA. Its two theorists were Burt Fried (my fellow graduate student) and Alfredo Banos (who had visited our group at the Livermore lab), and the chairman was Dave Saxon (whose work I had studied in my thesis research). Before accepting, Louise and I thought it prudent to try living in the LA area for a year. So I took a leave from the Livermore lab, and we moved to Sherman Oaks for the 1964-65 school year.

At UCLA, I taught the plasma theory course; among my students was George Johnston, who would be working with me some years later. My research there, stimulated by Al Wong's experiments, was with Burt and his postdoc Dave Sachs [28].

9. On the faculty at UCB, 1965-1971

While I was at UCLA, the Berkeley campus, in the throes of student riots, came up with a counter-offer, which I accepted. So in Fall 1965, I joined the regular faculty as a tenured Associate Professor, after five years as Lecturer. Part of my salary came from the Rad Lab, where I inherited Ted Northrop's office (he was moving to Goddard Space Flight Center), his desk (which I still have), and his thesis student Chuan Liu.

Ann Palm got us interested in the aurora, and Chuan proposed that he develop a theory of its dynamics for his thesis [29]. He was followed by Norman Albright [30] and Blake Patterson [31].

Ted provided me with a NASA grant to hire a postdoc, and I was fortunate to acquire Ed Frieman's prize student Ron Davidson in 1966. Ron's thesis was most impressive, and led to his publication output averaging one significant paper per month. Toward the end of his two-year

Journal of Physics: Conference Series **169** (2009) 012002 doi:10.1088/1742-6596/169/1/012002

stay, he presented a series of lectures on nonlinear plasma theory, which he then converted to his famous textbook.

The one paper that Ron and I co-authored, on weak turbulence [32], used a standard formula for wave entropy, derived by taking the classical limit of the formula for the entropy of a boson gas. The need for a quantum detour bothered me for years, and finally I discovered a purely classical derivation by Jaynes' maximum-entropy principle [33]. A few years later I met Rafi Levine, who told me of his prior use of Jaynes' method.

With Dave Baldwin, I developed a simple inverse method to determine the density profile of a plasma slab by external impedance measurements [34]. The extension to cylindrical geometry was mathematically challenging, so I referred that problem to Tony Degasperis [35].

When Toshio Nakayama became my postdoc in 1969, I proposed that we extend quasilinear theory from the well-studied spatially uniform case to the more challenging case of one-dimensional spatial variation [36]. This led to studying a cylindrical model of magnetized plasma [37], and finally to quasilinear diffusion in a tokamak [38]. That work utilized an action-angle formulation, based on Brian Taylor's guiding-center Lagrangian [39].

In September of 1970, we lost our home to a fire, set by an arsonist in dry grass on Fish Ranch Road. He was an employee of the regional park district, hired to fight fires, and paid by the fire, we were told. The strong dry winds spread the fire to our neighborhood, where a grove of eucalyptus, with loose bark and oily leaves, further spread the flames. Phone calls to the Oakland fire department were misinterpreted by their dispatcher, so their trucks came late. They requested help from Berkeley, whose hoses were of different diameter from Oakland's, and so did not fit the hydrants. Many volunteers, including students from the campus, came up the hills to assist the firemen. In all, 37 homes were totally destroyed; fortunately, no lives were lost.

Since the eucalyptus forest was now gone, we (wrongly) believed that our area would be safe from future fires, so we immediately decided to rebuild on-site. During the prior year, we had been remodeling our house, so that our children could have separate bedrooms. Thus our talented architects, Kolbeck and Christopherson, and builder, Al Heffley, were available to design and build our new home. We decided to take another leave at PPPL during Spring 1971, while our home was being rebuilt. In February 1971, we learned about the San Fernando earthquake, with its dreadful loss of life. Our architects' engineer went there to learn about reducing seismic risk, and advised appropriate strengthening of our incipient house.

10. 1971-1980

Shortly after our return in 1971, I acquired three new talented thesis students. To find suitable thesis projects for each of them, I turned to the prolific work of Marshall Rosenbluth. For Bruce Cohen, we examined Marshall's idea of beat heating, whereby the beat difference frequency of two lasers would be tuned to the local plasma frequency, and nonlinearly excite Langmuir waves, which would heat the plasma by Landau damping. We recognized that Marshall had overlooked the constraint of action conservation (Manley-Rowe relations), so that the primary effect would be the transfer of energy from the higher-frequency laser to the lower frequency laser. Joined by Ken Watson, we proposed instead a cascade process [40]. (At that time, conservation laws were not yet derived from variational principles and Noether's theorem. Instead, they were discovered by examining a set of coupled evolution equations, and so were often overlooked, being sensitive to relations among the nonlinear coupling coefficients.) Bruce examined the beat resonance in detail [41], including a simulation with my postdoc Claire Max [42].

For Dwight Nicholson, I proposed examining the question of absolute vs convective instability in three-wave interaction in a plasma gradient. This was a controversial and important question, as the convective case would saturate, while the absolute case would become fully nonlinear. Previous studies had used smooth background profiles, and led to convective growth resulting

Journal of Physics: Conference Series **169** (2009) 012002 doi:10.1088/1742-6596/169/1/012002

from phase relationships. We felt that a random perturbation of the density profile would disrupt the latter, and lead to absolute growth. This was confirmed in Dwight's study [43]. He then went on to show, in his thesis, that randomness was irrelevant, and that the same effect was produced by a sinusoidal perturbation [44].

For Mike Mostrom, I suggested the problem of Raman side-scatter instability, where the validity of WKB approximations made by Marshall and his co-workers was uncertain. His detailed analysis was incredibly complicated and sophisticated [45, 46].

By the time that Gary Smith joined us as my fourth student, I was cognizant of the newly discovered phenomenon of chaos. The idea that a deterministic system could exhibit random behavior was startling to the scientific community. I had been introduced to the subject in a colloquium at Princeton by Joe Ford, during my 1971 stay there. We read his papers (from 1969), studied the Henon-Heiles Hamiltonian (1964), were amazed by the revelations in Robert May's 1974 review in Science, and learned about Chirikov's resonance overlap studies. We had been unaware of Ed Lorenz' seminal paper of 1963, until pointed out by Ruelle & Takens.

At the time, there was discussion of the validity conditions for quasilinear diffusion. The standard derivation assumed a set of waves with neighboring phase velocities, and with random phases. Thus the randomness of diffusion would result from a random input. In contrast, I proposed that a nonrandom coherent set of neighboring resonances could produce diffusion, if the resonance islands overlapped. Such a set could be produced by the field of a single wave, with Doppler-shifted gyroharmonics. Gary tested this idea numerically, and it worked beautifully [47, 48, 49]. When I presented this work at the next APS meeting, it was met with incredulity. (When I told Tom Dupree that he had assumed such behavior in his own work, he said, "Yes, but I didn't believe it.") We searched for a term to describe this phenomenon, and came up with "intrinsic stochasticity"; not surprisingly, "chaos" was the term that stuck (Yorke 1975).

We presented this work at a major international conference in Cernobbio, in the summer of 1977: *Stochastic Behavior in Classical and Quantum Hamiltonian Systems* (VOLTA Memorial Conference, Como, organized by Giulio Casati). Here we first learned about quantum chaos, which would later become part of Steve McDonald's thesis [50].

In 1972, I proposed a new formulation of quasilinear theory, which had been plagued by the phenomenon of "fake diffusion", a second-order reversible change in the velocity distribution due to interaction of the wave field with nonresonant particles [51]. I showed that this effect represented the particle contribution to wave momentum and wave energy. This idea was clarified further and formalized by Bob Dewar, who introduced the concept of "oscillation center", and obtain its Hamiltonian dynamics by a canonical transformation [52].

In 1974, Hans Wilhelmsson convened an international workshop on plasma theory at Chalmers University in Goteborg, and in later years Nobel Symposia in Aspenasgarden. These meetings provided valuable interactions with Vladimir Karpman, Vadim Tsytovich, Lennart Stenflo, and many others. Lennart was a master of nonlinear wave interactions: we extended the Manley-Rowe action-conservation laws to the presence of a background high-frequency field [53] and we studied upper-hybrid solitons [54]. Later Lennart came to visit me at LBL, and produced an extension of 3-wave interaction to essential plasma nonuniformity [55].

Shayne Johnston became my postdoc in 1974, as a master of Bob Dewar's oscillation-center theory applied to nonlinear problems. We recognized that the widely-used heuristic ponderomotive potential was actually the oscillation-center Hamiltonian. We used these ideas to study nonlinear wave interactions [56], and, with George Johnston, we generalized this concept to allow for a ponderomotive vector-potential [57].

A major improvement in classical perturbation theory was the introduction of the powerful Lie transform, replacing the clumsy use of mixed-variable canonical transformations. We were made aware of it by Bob Dewar and by Alex Dragt and John Finn, and enthusiastically adopted it for all our studies [58]. By now I had three bright new students: John Cary, Robert Littlejohn, and Harry Mynick; we each produced a review of the Lie transform. John called his review "A Pedestrian's Guide", so Robert's became "An equestrian's guide", and Harry's was "for the jet set". My own review was personal and historical [59].

John's thesis research was devoted to developing the Lie transform, and applying it to a self-consistent nonlinear plasma-wave problem [60]. We discovered the remarkable identity between the linear susceptibility and the quadratic ponderomotive Hamiltonian, the K-χ theorem [61].

Harry Mynick chose his own thesis topics [62]: motivated by Mort Levine's experiment at LBL, one was a guiding-center dynamics when the gyro-orbit covers a large variation of magnetic field. Another topic was Harry's rigid-bar model for wave-trapped particles [63], based on a paper of O'Neil, Winfrey, & Malmberg.

For Robert Littlejohn, I selected an important problem whose solution had baffled me (as well as Bob Dewar) for some years: developing a rigorous and physically oriented Hamiltonian theory for guiding-center motion. By this time all three adiabatic invariants were known [64], Ted Northrop had published his comprehensive treatise using Hamiltonian ideas (1963), and yet a satisfactory systematic Hamiltonian theory was clearly needed. (I had used an unsatisfactory one in my 1972 quasilinear diffusion paper.) The sticking point was the universal belief among physicists (based on the standard textbooks) that a Hamiltonian formalism was necessarily based on canonically conjugate variables. The breakthrough came when Alan Weinstein gave us the galley proofs of his translation of Vladimir Arnold's new textbook on classical dynamics (published in 1978). Here was the exposition of symplectic geometry, and Hamiltonian dynamics with noncanonical Poisson brackets. Another new world opened for us. Robert was able to use these concepts to developed the needed guiding-center theory [65, 66, 67].

My next postdoc (1976-8) was Nino Pereira, an expert on solitons. With our visitor Lennart Stenflo, he generalized the nonlinear Schroedinger equation, to allow for soliton growth or damping [68]. With Gary, he studied chaos among trapped particles [69]. And with Ken's student Jim Meiss, he studied internal wave solitons, based on the Benjamin-Ono equation [70].

In 1978, my former student Chuan Liu (now a professor at Maryland) sent me his own student Celso Grebogi to postdoc with me. Our first project, with Robert, was to use the Lie-transform technique to extend the ponderomotive force formulas to magnetized plasma [71]. (The controversy was whether the gradient of the second-order potential acted also on the magnetic-field nonuniformity.) Then, with John Cary and Jim Meiss, with Henry Abarbanel and his student John David Crawford, and with Jin-Soo Kim (Andy Sessler's student), we began investigating various aspects of chaos [72, 73], as well as iterated maps for gyroresonance crossings. When Celso returned to Maryland in 1981, joining Ed Ott and Jim Yorke, he became eminent in the chaos field. (And a few weeks before my own Fest80, Louise and I journeyed to Aberdeen to celebrate his Fest60!)

Our entry into quantum chaos was stimulated by a paper of Wersinger, Ott, & Finn [74], examining the chaotic rays of lower-hybrid plasma waves. This led me to wonder what the corresponding wave field would look like. We needed a simple model to investigate the relation between chaotic rays and wave fields, or correspondingly, between classical orbits and quantum wave functions. Such a model was the stadium-shaped billiard, whose classical chaotic properties had recently been determined by Benettin & Strelcyn [75]. For a free particle, the corresponding quantum problem was the Helmholtz equation. A computer code for finding its eigenfunctions and eigenvalues had been written by our colleagues Bob Riddell & Joe Lepore, for the study of waves on Lake Tahoe [76]. My student Steve McDonald undertook a detailed study of the quantum billiard, and obtained its Wigner eigenvalue spectrum and its eigenfunctions' chaotic nodal lines. Our first paper [77] portrayed the world's first chaotic eigenfunction, and has become my most cited paper by far. In later work, Steve studied the Wigner function for a quantum chaotic billiard, testing Michael Berry's theories [78].

What an exciting decade, with all these new concepts to revel in: Manley-Rowe, classical and quantum chaos, oscillation center dynamics, Lie transform, noncanonical Hamiltonian dynamics for guiding-centers. To master them, we had a weekly dynamics seminar, in addition to our long-standing plasma seminar. We worked our way through important review articles and textbooks, especially Arnold's two and later Sinai *et al.* on ergodic theory.

11. 1980-1987

Steve had taken a math course from Andrew Majda, on pseudodifferential operators. He then developed the Wigner-Weyl symbol calculus as a tool for plasma theory. Our first important result was a derivation (as opposed to a statement) of the linear wave kinetic equation for the action density in ray phase space [79]. With Celso, we then combined the wave evolution with that of the oscillation centers [80], including resonant interaction. Steve later prepared an important and extensive review paper on the symbol calculus [81].

In late 1981, Bob Dewar sent me his student Eliezer Rosengaus to be my next postdoc. At the Quantum Chaos Conference in Como, in 1983, we presented our ideas on classical wave chaos.

Shortly after Robert's discovery of the Poisson brackets for guiding-center dynamics, Phil Morrison & John Greene [82] presented a noncanonical Hamiltonian formulation of MHD. With my student Rich Spencer, we applied these ideas to the two-fluid model of plasma dynamics [83]. After Phil found a Hamiltonian structure for the Vlasov equation [84], I worked with him to find such a non-Hamiltonian geometric structure for the quasilinear equations [85]. Such concerns led me to a dissipative bracket formalism [86], found independently and simultaneously by Phil [87] and by Miroslav Grmela [88]. I then generalized the formalism to a covariant form [89], and further with Lukasz Turski [90].

In 1983, Steve Omohundro became my thesis student with broad interests in plasma physics and in math, so Alan Weinstein was his co-advisor. To a large extent he developed his own ideas, and published his thesis [91] as a book.

In 1983 Philippe Similon became my postdoc. One of our first projects concerned the stabilization of an unstable plasma by the ponderomotive effects of an applied rf wave, which had been demonstrated by Noah Hershkowitz' group at Madison. From my work with John Cary, Shayne Johnston, & George Johnston, we expected a number of new subtle effects, which Philippe obtained by formulating the problem by a variational principle [92, 93, 94]. This success confirmed my gradually growing conviction of the power of variational formulations. Earlier evidence of this power had been Robert's Lagrangian reformulation of guiding-center dynamics, Herb Berk & Rene Dominguez' variational approach to systematic approximations, and the pioneering work of Fred Crawford, John Dougherty, and Bob Dewar, demonstrating Noether's method for obtaining conservation laws from continuous symmetries. Robert had converted the Lie transform to a Lagrangian setting, working with John Cary. Building on these ideas, Darryl Holm and I derived a variational (and relativistically covariant) approach to the dynamics of an unmagnetized oscillation-center plasma interacting nonresonantly with an eikonal wave [95]. The needed extension to a magnetized plasma was a major undertaking, so I assigned it as a thesis topic to my student Bruce Boghosian. The first steps in this project were developed by Philippe, who produced a covariant guiding-center plasma dynamics, with a complete set of conservation laws [96]. Then Bruce extended the formulation to the covariant oscillation-center plasma, interacting ponderomotively with an eikonal wave. (His important thesis was never converted to a journal publication, but is now available on the Web [97].) The next extension, to include resonant interaction, and the consequent reaction of the background fields, was accomplished by Huanchun Ye in his thesis [98, 99].

In 1985, a semester-long workshop on plasma theory was held at the Santa Barbara Institute for Theoretical Physics, organized by John Krommes, Tom O'Neil, and Ed Ott. One of my projects was to work with Celso, to complete what Steve McDonald and I had begun on obtaining coupled equations for interacting particles and waves [80]. The other was suggested by Carl Oberman, who proposed using the methods of my group (variational principles and Lie transforms) to deal with a problem that John David Crawford had formulated, the coupling of modes by beat resonance. After we had finished, I discovered to my delight that Shayne had done the same calculation in unpublished notes in my files [100].

Huanchun Ye started as my student in 1985. My first project for him was to apply our Lie transform methods to determine a three-wave coupling coefficient under kinetic (as opposed to fluid-model) conditions. As a comparison, we had the mysterious (to us) powerful methods developed by Lennart Stenflo and his former student (now colleague at Umea) Jonas Larsson. Their approach led to an expression with six terms, while ours had 192 terms, as I recall. More about Jonas later.

12. Mode conversion, 1987 on

In 1986, Lazar Friedland came to spend a sabbatical year working with me. I had invited him after seeing his incisive approach to linear mode conversion, presented at an APS meeting. The first of his chief ideas was a systematic reduction of a multicomponent wave equation to a two-component form, representing the two waves involved in the conversion process. After linearizing the two wave operators with respect to their x,k dependence (and treating the coupling as a constant), he cleverly solved the equations for the transmission coefficient of the incident wave. Examining his expression, I recognized a Poisson bracket hiding there; so I felt that there must be a far simpler and more fundamental approach to this problem. My main tool was the metaplectic transform, a powerful generalization of the Fourier transform, which Robert had recently developed. This was the field/operator transformation representing an arbitrary linear canonical transformation of (x,k) phase space. Indeed this approach made the solution of the coupled equations trivial, as they became a first-order ODE.

Toward the end of Lazar's stay, I returned to Chicago to look after my mother during her terminal illness. She passed away at age 97.

In 1987, Walt Sadowski convened a Science Court, to provide guidance on the problem of rf heating in the ion gyrofrequency range for tokamak plasmas. One team, called the "defense", presented their approach, while another team, called the "prosecutor", critiqued it. The three "judges" (myself, Charles Karney, and Ira Bernstein as chief justice) then prepared a report with recommendations. Reading through the literature, I realized that the mode-conversion analyses that Lazar and I had been doing as an academic study were immediately applicable to a practical problem in fusion energy.

Accordingly, Huanchun and I turned our attention to linear conversion problems. Our first example was the linear ion-cyclotron echo (which we decided not to abbreviate as LICE), based on a fluid model (including the anisotropic pressure tensor) which Lazar had introduced. We showed that the standard Budden model represented a two-step conversion, with the Budden resonance actually an eikonal wave at fixed x, propagating only in k-space.

After our publications [101, 102], Tudor Johnston informed us that the delayed echo idea had been discussed by Muldrew years before, in 1969.

In August 1989, two visitors joined our group for a year: Maria Ekiel-Jezewska from Warsaw & Tor Fla from Tromso. Maria & Tor derived a generalized vector nonlinear Schroedinger equation, with resonant particle effects included. In addition Tor joined a number of our other projects, including multiple mode conversion [103].

Alain Brizard began his collaboration with me in October 1989, arriving, as my new postdoc, two days before the Loma Prieta earthquake. His Princeton thesis had demonstrated his expertise in Lie-transform and variational methods. Since then, we have worked together on many problems associated with mode conversion and with variational principles, publishing 13 papers as co-authors, so far.

At William & Mary, Gene Tracy started his collaboration with me at about the same time, beginning with a generalization of mode conversion theory to incoherent waves [104]. He then put my heuristic analysis on a sound mathematical footing [105]. Over the years, Gene, Alain and I have been kindred spirits, with a remarkably similar approach and philosophy about plasma physics.

In July 1990, a conference on "nonlinear and chaotic phenomena" was held in Edmonton, organized by Rozmus and Tuszynski. Here I presented an extensive review of our group's work on "linear and nonchaotic phenomena", with historical details.

In October 1991, fire struck the Berkeley-Oakland Hills again. On a Saturday, there had been a minor fire, caused by debris from a remodeling project near our home. After putting it out, the Oakland Fire Department left the scene, with embers still hot. The next morning, strong winds revived the fire, which was quickly out of control, spreading as a fire-storm. Some 25 lives were lost, and 3,000 homes. We now knew it was futile to rebuild, so purchased a house in Orinda. For a week we stayed with Wulf & Erika Kunkel, and then for a month with Don & Lynn Glaser. While there, we were horrified to learn that Dwight Nicholson and colleagues had been murdered. Recognizing Dwight's exceptional qualities of character, Marty Goldman & Herb Berk initiated an APS medal in his memory, to be awarded annually.

Dan Cook was my last thesis student, his topic being the ray-based analysis of minority heating by mode conversion [106]. Lazar had shown that Landau resonance and gyroresonance could be quantitatively interpreted as conversion to ballistic waves. In minority heating, there is a combination of conversion to a collective wave (a minority quasi-Bernstein wave) as well as Doppler-broadened gyroresonance (interpreted as conversion to a continuum of Case-van Kampen modes). Accordingly, it was necessary to incorporate the (damped) collective wave within the complete set of eigenfunctions of the linearized Vlasov equation. This was accomplished by analytic continuation to complex-valued velocity, the technique developed by John David Crawford and Peter Hislop. The reflection of the incident magnetosonic wave then resulted from a second conversion process [107]. The analysis was then extended from the slab model to two-dimensionality [108].

Jonas Larsson, a frequent visitor, introduced a number of radical new ideas into plasma theory, which we appreciated and utilized. One was the concept of pseudo-Hilbert space, and the corresponding idea of a pseudo-Hermitian operator. (The former has an inner product with a nonpositive norm.) Jonas showed that the linearization of any Hamiltonian produces a Hermitian or pseudo-Hermitian operator. Since the Maxwell-Vlasov and Poisson-Vlasov systems are examples of this, their linearization can then directly be represented by a quadratic variational principle. This enabled us to extend action-conservation laws to non-eikonal situations [109].

My last postdoc was Yi-Ming Liang, with whom Gene and I initiated a study of wave emission as a conversion process [110].

After Robert Littlejohn was appointed to our faculty in 1983 and joined our research group at LBL, he continued providing powerful new tools for our research, such as the metaplectic transform. In addition, his thesis students often showed great interest in our studies, and made valuable contributions. Chief among them was Jim Morehead, who continued as a postdoc. He provided crucial insight in our work on multiple mode conversion [103], and was a valuable co-worker in some five other papers. Also, Robert's students Greg Flynn [111] and Yukkei Hui [112] joined our studies. And Kevin Mitchell patiently taught me how to use my computer.

In 1997, Darryl suggested that our mode-conversion methodology might be transferable to ocean waves, which he had been studying at LANL. In Philander's treatise on El Nino, we found a beautiful diagram of a mode-conversion process (an avoided-crossing dispersion diagram), from work by Cane & Sarachik [113]. We proceeded to apply our methods to the problem of a spatially variable background (the depth of the thermocline), and published our first oceanography paper [114], involving energy transfer from an equatorially trapped Rossby-Kelvin wave to a coastally trapped Kelvin wave. Later, Brad Shadwick produced simulation movies, to test our theoretical approximations. Our work was brought to the attention of oceanographers by Vanneste [115], and then applied by Rene Tailleux to Rossby waves [116].

Having become interested in geophysical fluid dynamics, I gave a physics course on the subject, using Cushman-Roisin's excellent textbook. We then became aware of Eckart resonance [117], involving mode conversion among internal waves in the ocean. A similar resonance occurs among Doppler-ducted internal waves in the atmosphere, studied by Dave Fritts [118]. To deal with linearization about an atmosphere with sheared flow, we needed Jonas' generalization of Hamiltonian dynamics [119]. He had trained his graduate student Krister Wiklund, whom he sent to work with me on this problem [120].

Andre Jaun, a Swiss computational plasma physicist working in Stockholm on full-wave modeling, recognized the value of our ray-based methods. In 2000, he joined our team, proposing the implementation of our approach for realistic tokamak geometries.

I retired from teaching in 1998, having started in 1959. Preparing my lectures and homework assignments was a wonderful stimulus to deeper understanding. Each time I taught a course, I would try to dig deeper. Without teaching, my research would surely have suffered.

13. Colleagues

Wulf Kunkel was group leader of the plasma program at LBL from 1971 to 1991, and was my plasma colleague on the Physics Department faculty. Over these many years, this good friend was a continual source of valuable advice, as we discussed problems in physics and in the real world of people.

Andy Sessler was my theoretical colleague for many years at LBL, and director for part of that time. As a leading accelerator theorist, he called my attention to Chirikov's pioneering work on chaos. His students Jim Hammer, Paul Channell, Jonathan Wurtele, & Jin-Soo Kim were active members of our group.

In our campus EE department, Ned Birdsall trained all my students in the plasma simulation techniques which he and Bruce Langdon pioneered. One homework project for Dwight Nicholson, the motion of a charged particle in an electrostatic wave packet, led us to the concept of oscillation-center velocity. His colleagues Allan Lichtenberg and Mike Lieberman taught a valuable course on MHD, and wrote a definitive textbook on Hamiltonian dynamics.

Oscar Manley for many years provided our grant support from Washington, both in fusion and in basic energy science. He was most remarkable in his deep understanding of our research, based on his strong physics and math background. After retirement he co-authored incisive work on turbulence, with Foias and Temam. His mathematician friend Yvain Treve helped us make valuable contact with the math community, in particular with Alan Weinstein and Jerry Marsden.

Over the years, short-term visitors provided valuable insights and knowledge. Chief among them were John Krommes, with his sophisticated approach to turbulence and broad understanding of plasma theory, and Jonas Larsson, who introduced powerful new mathematical tools.

I feel very fortunate to have found a career in a field where I could grow with the field itself. I have had exceptionally bright students, congenial stimulating supportive colleagues, and a stable and secure institutional home. My wonderful family have fulfilled my life: my wife Louise, our son Joel, our daughter Janet, our son-in-law Tom, and our grandsons Alexander and Aaron.

It has truly been a happy time.

References

[1] P. Kollander, *Frederick III: Germany's Liberal Emperor* (Greenwood, 1995).
[2] Max Grinnell, *Hyde Park* (Arcadia, 2001).

[3] P. M. Ascoli, *Julius Rosenwald* (Indiana, 2006).
[4] J. L. Liebman, *Piece of Mind* (Simon & Schuster, 1946).
[5] J. J. Weinstein, *Advocate of the People* (Ktav, 1980).
[6] W. McNeill, *Hutchins' University* (Chicago, 1991).
[7] J. Orear, A. Rosenfeld, R. Schluter, *Nuclear Physics: a course given by Enrico Fermi* (Chicago, 1950).
[8] L. Teng & J. Tuck, U. S. Patent 2812463 (1951).
[9] R. Courant & K. Friedrichs, *Supersonic Flow and Shock Waves* (Interscience, 1948).
[10] L. Spitzer, *Physics of Fully Ionized Gases* (Interscience, 1956).
[11] M. N. Rosenbluth and A. N. Kaufman, Phys. Rev. **109** (1958) 1.
[12] A. N. Kaufman, Phys. Fluids **1** (1958) 252.
[13] A. N. Kaufman, Phys. Fluids **3** (1960) 610.
[14] S. Chandrasekhar, A. N. Kaufman, and K. M. Watson, Proc. Roy. Soc. **245** (1958) 435.
[15] A. N. Kaufman and K. M. Watson, Phys. Fluids **4** (1961) 655.
[16] A. N. Kaufman and K. M. Watson, Phys. Fluids **4** (1961) 931.
[17] A. N. Kaufman and S. Toshio, Phys. Rev. A **136** (1964) 1614.
[18] A. N. Kaufman and S. Toshio, J. Chem. Phys. **37** (1962) 1988.
[19] A. N. Kaufman and P. S. Rostler, Phys. Fluids **14** (1971) 446.
[20] A. N. Kaufman, Am. J. Phys. **29** (1961) 626.
[21] A. N. Kaufman, Ann. Phys. **18** (1962) 264.
[22] A. N. Kaufman, Phys. Fluids **8** (1965) 935.
[23] W. Kunkel, ed., *Plasma Physics in Theory and Application* (McGraw-Hill, 1966).
[24] L. Klauder, *Statistical Theory of Nonlinear Effects in the Polarization of an Imperfect Gas* (1964).
[25] S. Fisher, *Stability and Oscillation Analysis of Magnetic Mirror Systems with Conducting End Plates* (1963).
[26] G. Pearson, *The Effect of Wave-Particle Interactions on the Stability of a Current-Carrying Plasma* (1965).
[27] J. C. Price, *Plasma Kinetic Equations* (1966).
[28] B. D. Fried, A. N. Kaufman, and D. L. Sachs, Phys. Fluids **9** (1966) 292.
[29] C. Liu, *Low Energy Electrostatic Instabilities in the Magnetosphere* (1968).
[30] N. Albright, *Quasilinear Evolution of the Transverse Instability Driven by Temperature Anisotropy of an Unmagnetized Plasma* (1969).
[31] B. Patterson, *Convective Evolution of Large Amplitude Alfven Waves in the Solar Wind* (1971).
[32] R. C. Davidson and A. N. Kaufman, J. Plasma Phys. **3** (1969) 97.
[33] A. N. Kaufman, Phys. Fluids **29** (1986) 2326.
[34] D. E. Baldwin and A. N. Kaufman, Phys. Fluids **12** (1969) 1526.
[35] A. Degasperis, J. Math. Phys. **11** (1970) 3392.
[36] A. N. Kaufman and T. Nakayama, Phys. Fluids **13** (1970) 956.
[37] A. N. Kaufman, Phys. Fluids **14** (1971) 387.
[38] A. N. Kaufman, Phys. Fluids **15** (1972) 1063.
[39] J. B. Taylor, Phys. Fluids **7** (1964) 767.
[40] B. I. Cohen, A. N. Kaufman, and K. M. Watson, Phys. Rev. Lett. **29** (1972) 581.
[41] B. I. Cohen, *Theoretical Studies of Some Nonlinear Laser-Plasma Interactions* (1975).
[42] B. I. Cohen, M. A. Mostrom, D. R. Nicholson, A. N. Kaufman, C. E. Max, and A. B. Langdon, Phys. Fluids **18** (1975) 470.
[43] D. R. Nicholson and A. N. Kaufman, Phys. Rev. Lett. **33** (1974) 1207.
[44] D. R. Nicholson, *Parametric Instabilities in Inhomogeneous Plasma* (1975).
[45] M. A. Mostrom, *Raman Side-Scatter Instability in Nonuniform Plasma* (1978).
[46] M. A. Mostrom and A. N. Kaufman, Phys. Rev. Lett. **42** (1979) 644.
[47] G. R. Smith and A. N. Kaufman, Phys. Rev. Lett. **34** (1975) 1613.
[48] G. R. Smith, *Stochastic Acceleration by a Single Wave in a Magnetized Plasma* (1977).
[49] G. R. Smith and A. N. Kaufman, Phys. Fluids **21** (1978) 2230.
[50] S. W. MacDonald, *Wave Dynamics of Regular and Chaotic Rays* (1983).
[51] A. N. Kaufman, J. Plasma Phys. **8** (1972) 1.
[52] R. L. Dewar, Phys. Fluids **16** (1973) 1102.
[53] A. N. Kaufman and L. Stenflo, Plasma Phys. **17** (1975) 403.
[54] A. N. Kaufman and L. Stenflo, Phys. Scrip. **11** (1975) 269.
[55] A. N. Kaufman and L. Stenflo, Phys. Scrip. **19** (1979) 523.
[56] S. Johnston and A. N. Kaufman, J. Plasma Phys. **22** (1979) 105.
[57] S. Johnston, A. N. Kaufman, and G. L. Johnston, J. Plasma Phys. **20** (1978) 365.
[58] S. Johnston and A. N. Kaufman, Phys. Rev. Lett. **40** (1978) 1266.
[59] A. N. Kaufman, *The Lie transform: A new approach to classical perturbation theory*, in AIP Conference

Proceedings **46** (1978) 286.
[60] J. R. Cary, *Nonlinear Wave Evolution in Vlasov Plasma: A Lie-Transform Analysis* (1979).
[61] J. R. Cary and A. N. Kaufman, Phys. Rev. Lett. **39** (1977) 402.
[62] H. E. Mynick, *Equilibrium and Stability in Strongly Inhomogeneous Plasmas* (1978).
[63] H. E. Mynick and A. N. Kaufman, Phys. Fluids **21** (1978) 653.
[64] T. G. Northrop and E. Teller, Phys. Rev. **117** (1960) 215.
[65] R. G. Littlejohn, J. Math. Phys. **20** (1979) 2445.
[66] R. G. Littlejohn, *Hamiltonian Theory of Guiding Center Motion* (1980).
[67] R. G. Littlejohn, Phys. Fluids **24** (1981) 1730.
[68] N. R. Pereira and L. Stenflo, Phys. Fluids **20** (1977) 1733.
[69] G. R. Smith and N. R. Pereira, Phys. Fluids **21** (1978) 2253.
[70] J. D. Meiss and N. R. Pereira, Phys. Fluids **21** (1978) 700.
[71] C. Grebogi, A. N. Kaufman, and R. G. Littlejohn, Phys. Rev. Lett. **43** (1979) 1668.
[72] C. Grebogi and A. N. Kaufman, Phys. Rev. A **24** (1981) 2829.
[73] J. D. Meiss, J. R. Cary, C. Grebogi, J. D. Crawford, A. N. Kaufman, and H. D. I. Abarbanel, Physica D **6** (1983) 375.
[74] J.-M. Wersinger, E. Ott, and J. M. Finn, Phys. Fluids **21** (1978) 2263.
[75] G. Benettin and J.-M. Strelcyn, Phys. Rev. A **17** (1978) 773.
[76] Bob Riddell & Joe Lepore, *Boundary Dipole Distributions for the Solution of Helmholtz Equations*, LBL-3036 (1974).
[77] S. W. MacDonald and A. N. Kaufman, Phys. Rev. Lett. **42** (1979) 1189.
[78] S. W. MacDonald and A. N. Kaufman, Phys. Rev. A **37** (1988) 3067.
[79] S. W. MacDonald and A. N. Kaufman, Phys. Rev. A **32** (1985) 1708.
[80] S. W. MacDonald, C. Grebogi, and A. N. Kaufman, Phys. Lett. A **111** (1985) 19.
[81] S. W. MacDonald, Phys. Rep. **158** (1988) 337.
[82] P. J. Morrison and J. M. Greene, Phys. Rev. Lett. **45** (1980) 790.
[83] R. G. Spencer and A. N. Kaufman, Phys. Rev. A **25** (1982) 2437.
[84] P. J. Morrison, Phys. Lett. A **80** (1980) 383.
[85] A. N. Kaufman and P. J. Morrison, Phys. Lett. A **88** (1982) 405.
[86] A. N. Kaufman, Phys. Lett. A **100** (1984) 419.
[87] P. J. Morrison, Phys. Lett. A **100** (1984) 423.
[88] M. Grmela, Phys. Lett. A **102** (1984) 355.
[89] A. N. Kaufman, Phys. Lett. A **109** (1985) 87.
[90] L. A. Turski and A. N. Kaufman, Phys. Lett. A **120** (1987) 331.
[91] S. M. Omohundro, *Geometric Perturbation Theory & Plasma Physics* (1985).
[92] P. L. Similon and A. N. Kaufman, Phys. Rev. Lett. **53** (1984) 1061.
[93] P. L. Similon, A. N. Kaufman, and D. D. Holm, Phys. Lett. A **106** (1984) 29.
[94] P. L. Similon, A. N. Kaufman, and D. D. Holm, Phys. Fluids **29** (1986) 1908.
[95] A. N. Kaufman and D. D. Holm, Phys. Lett. A **105** (1984) 277.
[96] P. L. Similon, Phys. Lett. A **112** (1985) 33.
[97] B. M. Boghosian, *Covariant Lagrangian Methods of Relativistic Plasma Theory* (1987); see arXiv:physics/0307148.
[98] H. Ye, *Wave Dynamics in Phase Space & Ion Gyroresonant Absorption* (1989).
[99] H. Ye and A. N. Kaufman, Phys. Fluids B **4** (1992) 1735.
[100] J. D. Crawford, S. Johnston, A. N. Kaufman, and C. Oberman, Phys. Fluids **29** (1986) 3219; Erratum, Phys. Fluids **30** (1987) 618.
[101] H. Ye and A. N. Kaufman, Phys. Rev. Lett. **60** (1988) 1642.
[102] H. Ye and A. N. Kaufman, Phys. Rev. Lett. **61** (1988) 2762.
[103] Y.-M. Liang, J. J. Morehead, D. R. Cook, T. Fla, and A. N. Kaufman, Phys. Lett. A **193** (1994) 82.
[104] E. R. Tracy and A. N. Kaufman, Phys. Rev. Lett. **64** (1990) 1621.
[105] E. R. Tracy and A. N. Kaufman, Phys. Rev. E **48** (1993) 2196.
[106] D. R. Cook, *Wave Conversion in Phase Space & Plasma Gyroresonance* (1993).
[107] D. R. Cook, A. N. Kaufman, E. R. Tracy, and T. Fla, Phys. Lett. A **175** (1993) 326.
[108] D. R. Cook, A. N. Kaufman, A. J. Brizard, H. Ye, E. R. Tracy, Phys. Lett. A **178** (1993) 413.
[109] A. J. Brizard, D. R. Cook, and A. N. Kaufman, Phys. Rev. Lett. **70** (1993) 521.
[110] E. R. Tracy, A. N. Kaufman, and Y.-M. Liang, Phys. Plasmas **2** (1995) 4413.
[111] D. R. Cook, W. G. Flynn, J J. Morehead, and A. N. Kaufman, Phys. Lett. A **174** (1993) 53.
[112] A. N. Kaufman, H. Ye, and Y. Hui, Phys. Lett. A **120** (1987) 327.
[113] M. A. Cane and E. S. Sarachik, J. Marine Res. **37** (1979) 355.

[114] A. N. Kaufman, J. J. Morehead, A. J. Brizard, and E. R. Tracy, J. Fluid Mech. **394** (1999) 175.
[115] J. Vanneste, J. Phys. Oceanogr. **31** (2001) 1922.
[116] R. Tailleux and J. C. McWilliams, J. Fluid Mech. **473** (2002) 295.
[117] C. Eckart, Phys. Fluids **4** (1961) 791.
[118] D. Fritts and L. Yuan, J. Geophys. Res. **94** (1989) 18455.
[119] J. Larsson, J. Plasma Phys. **55** (1996) 235.
[120] K. Wiklund and A. N. Kaufman, Phys. Lett. A **279** (2001) 67.

Plasma Theory, Wave Kinetics, Nonlinear Dynamics: KaufmanFest 2007 IOP Publishing
Journal of Physics: Conference Series **169** (2009) 012003 doi:10.1088/1742-6596/169/1/012003

Variational Principles for Reduced Plasma Physics

Alain J. Brizard
Department of Physics, Saint Michael's College, Colchester, VT 05439, USA

Abstract. Reduced equations that describe low-frequency plasma dynamics play an important role in our understanding of plasma behavior over long time scales. One of the oldest paradigms for reduced plasma dynamics involves the ponderomotive Hamiltonian formulation of the oscillation-center dynamics of charged particles (over slow space-time scales) in a weakly-nonuniform background plasma perturbed by an electromagnetic field with fast space-time scales. These reduced plasma equations are derived here by Lie-transform and variational methods for the case of a weakly-magnetized background plasma. In particular, both methods are used to derive explicit expressions for the ponderomotive polarization and magnetization, which appear in the oscillation-center Vlasov-Maxwell equations.

1. Introduction

Our understanding of complex plasma phenomena can be significantly improved by the asymptotic elimination (or decoupling) of fast space-time scales from slow space-time scales of interest. Over the past 30 years, Lie-transform perturbation methods [1-3] have played an important role in the process of dynamical reduction in plasma physics. Standard examples of reduced plasma dynamics include the elimination of fast space-time scales associated with the gyromotion of a charged particle in a strong magnetic field (i.e., the *guiding-center* reduction [4-6] and the *gyrocenter* reduction [7-8]) or the elimination of the fast eikonal space-time scales associated with charged-particle motion perturbed by a high-frequency, short-wavelength electromagnetic wave (the *oscillation-center* reduction [9-18]) in a weakly-nonuniform plasma.

Note that, while the process of dynamical reduction is implicitly associated with the elimination of fast space-time scales from a set of dynamical equations, the process may or may not explicitly reduce the number of degrees of freedom. When it does (e.g., guiding-center reduction), a fast orbital angle (used to describe particle motion) becomes asymptotically ignorable and its conjugate action becomes an adiabatic invariant. Whether or not the number of degrees of freedom is reduced, however, the numerical integration of these dynamically-reduced equations of motion over long space-time scales of interest, which generally allows for more realistic plasma geometries to be considered, presents an important practical application of dynamical reduction in plasma physics.

Variational formulations of reduced plasma dynamics exist for single-particle Hamiltonian dynamics as well as Vlasov-Maxwell (kinetic) and fluid self-consistent theories. First, the variational formulation for reduced single-particle dynamics relies on the dynamical reduction of the phase-space Lagrangian by Lie-transform methods [3]. The reduced single-particle dynamics is expressed in terms of a reduced Hamiltonian and a reduced Poisson-bracket structure (derived from the symplectic part of the reduced phase-space Lagrangian). Conservation laws are associated with the invariance of the reduced phase-space Lagrangian and Hamiltonian on fast degrees of freedom (e.g., ignorable or fast angles) and the corresponding invariants (e.g., exact or adiabatic actions) appear explicitly in the symplectic part of the reduced phase-space Lagrangian. Second, the variational formulation of the reduced Vlasov-Maxwell (kinetic) equations, on the other hand, is intimately connected to the variational formulation of reduced single-particle dynamics through a constrained variational principle in extended phase space [19]. Conservation laws for the reduced Vlasov-Maxwell equations are derived by applying the Noether method on the reduced Lagrangian density. Third, the variational formulation for reduced fluid dynamics relies on the existence of the reduced fluid Lagrangian density, from which exact conservation laws are derived once again by the Noether method [20].

1.1. Berkeley School of Plasma Physics

In all of these developments, the Berkeley School of Plasma Physics, led by Allan N. Kaufman in collaboration with his graduate students and postdocs, has played a fundamental role. Allan's pioneering work in plasma theory over 50 years (with over 100 papers), which is reviewed in his memoirs (appearing in this volume), can be divided into three (approximately) separate periods. During the first period (from 1957 to 1969), Allan's work focussed on plasma dynamics in a magnetic field (in collaborations with S. Chandrasekhar and K. M. Watson) and the statistical physics of an imperfect gas. During the second period (from 1970 to 1987), Allan's work focussed on nonlinear plasma waves, the Hamiltonian formulation of plasma physics (which included the Lie-transform perturbation analysis of oscillation-center and guiding-center dynamics), the construction of dissipative Poisson brackets (by combining an entropy-conserving Poisson bracket and an energy-conserving dissipative bracket), wave & particle chaos, and wave kinetic theory. During the third period (from 1987 to the present), Allan's work focussed on linear wave conversion in plasmas and fluids, as well as the non-eikonal formulation of wave-action conservation laws in plasma physics.

The main characteristics of the Berkeley School of plasma physics, which were developed mainly during the second and third periods of Allan's work, involve the development and applications of Hamiltonian and Lagrangian methods in plasma physics. In fact, with a few exceptions (e.g., R. L. Dewar [21], P. J. Morrison [22], and J. A. Krommes [23]), the use of these methods in plasma physics is intimately associated with Allan's name. It is worth mentioning that the Berkeley School has also greatly benefited from the influx of many postdocs (from Princeton's PPL and elsewhere), visitors, and collaborators (see Allan's memoirs for details).

The development and applications of Hamiltonian and Lagrangian methods in linear wave conversion (during the third period) are reviewed by Tracy and Brizard in a separate paper in this volume. Here, I present the derivation of the oscillation-center dynamics of charged particles (developed during the second period) in a weakly-magnetized background plasma perturbed by high-frequency electromagnetic field fluctuations. The emphasis will be placed on applications of Lie-transform perturbation methods (with John Cary [2] and Robert Littlejohn [3], two of Allan's graduate students, acting as major participants in their developments) and applications of variational (action) principles to the development of the ponderomotive Hamiltonian oscillation-center theory (with Allan's students John Cary [11] and Bruce Boghosian [17] as well as a series of Allan's postdocs including Shayne Johnston [9], Celso Grebogi [10], and Philippe Similon [14-16] acting as major participants in their developments). The purpose of the paper is not to present a complete survey of the historical development of oscillation-center Hamiltonian dynamics in plasma physics (which is briefly discussed in Allan's memoirs). Instead I wish to present a new (and hopefully interesting) application of Lie-transform perturbation and variational methods in plasma physics.

In the present paper, the ponderomotive (eikonal-averaged) polarization and magnetization effects due to a high-frequency electromagnetic wave propagating in a weakly-magnetized plasma are investigated by two complementary methods. The ponderomotive polarization and magnetization are derived either from an oscillation-center variational principle involving the weak background electromagnetic fields $(\mathbf{E}_0, \mathbf{B}_0)$ as variational fields, or from the (Lie-transform) push-forward relation between the particle fluid moments and the oscillation-center fluid moments. An important consequence of the variational formulation is that an exact energy-momentum conservation law, which is derived by the Noether method, is expressed explicitly in terms of linear and ponderomotive polarization and magnetization. We note that while some of the fundamental issues raised by this work are resolved within a covariant relativistic treatment, we present a non-relativistic treatment here (except when discussing the moving-magnetic-dipole controbution to polarization in Sec. 3.4) and postpone the relativistic treatment for future work.

1.2. Organization

The remainder of the paper is organized as follows. In Sec. 2, the variational formulation of the exact Vlasov-Maxwell equations is reviewed [19]. The formulation is given in terms of extended phase-space coordinates, in which the energy-time canonical-coordinate pair is added to the six-dimensional regular phase-space coordinates. From the Vlasov-Maxwell action functional, we also derive conservation laws by using the Noether method. In Sec. 3, the variational formulation of a generic set of reduced Vlasov-Maxwell equations is presented [24]. Here, the terms "exact" and "reduced" are used to emphasize the fact that, while the exact Vlasov-Maxwell equations exhibit the full range of space-time scales associated with exact particle Hamiltonian dynamics, the reduced Vlasov-Maxwell equations exhibit only the long space-time scales associated with reduced Hamiltonian dynamics. The explicit derivations of general expressions for the polarization and magnetization by push-forward (Lie-transform) method and by variational method are presented. The energy-momentum conservation laws for the self-consistent reduced Vlasov-Maxwell equations are also derived by the Noether method. In Sec. 4, the Lie-transform perturbation analysis of oscillation-center Hamiltonian dynamics is presented in terms of non-canonical coordinates, so that both the Poisson-bracket (symplectic) structure and the Hamiltonian are perturbed by the electromagnetic wave fields. Next, the linear and nonlinear polarization and magnetization are derived by variational and push-forward methods. Lastly, a summary of this work is presented in Sec. 5.

2. Exact Vlasov-Maxwell Equations

2.1. Extended Hamiltonian Dynamics

The most general setting for Hamiltonian perturbation theory [24] is the eight-dimensional extended phase space with coordinates $z^a = (t, \mathbf{x}; w, \mathbf{p})$, where w and $\mathbf{p}$ denote the kinetic energy-momentum coordinates and $(\mathbf{x}, t)$ denote the space-time location of a charged particle (of mass m and charge e). Here, the Hamiltonian description of charged-particle dynamics in an electromagnetic field $\mathsf{F}_{\mu\nu} \equiv \partial_\mu A_\nu - \partial_\nu A_\mu$, represented by the four-potential $A_\mu \equiv (-\Phi, \mathbf{A})$ [using the space-time metric $g^{\mu\nu} = \mathrm{diag}(-1, 1, 1, 1)$], is expressed in extended phase space in terms of (i) the extended Hamiltonian

$$\mathcal{H}(\mathsf{z}) = \frac{1}{2m} |\mathbf{p}|^2 - w \equiv 0, \tag{1}$$

and (ii) the extended phase-space Lagrangian (summation over repeated indices is implied)

$$\Gamma = \left(\mathbf{p} + \frac{e}{c}\mathbf{A}\right) \cdot \mathsf{d}\mathbf{x} - (w + e\,\Phi)\,\mathsf{d}t = \left(p_\mu + \frac{e}{c}A_\mu\right)\mathsf{d}x^\mu \equiv \Gamma_a\,\mathsf{d}z^a, \tag{2}$$

where $x^\mu \equiv (ct, \mathbf{x})$ and $p_\mu \equiv (-w/c, \mathbf{p})$. The constraint $\mathcal{H} = 0$ in Eq. (1) implies that the physical motion takes place on the surface $w = |\mathbf{p}|^2/2m$.

The extended equations of motion are obtained from the variational principle

$$0 = \delta \int \left(\Gamma - \mathcal{H}\, d\tau \right), \tag{3}$$

where τ is the Hamiltonian orbit parameter in extended phase space. The variational principle (3) yields the Euler-Lagrange equations

$$\omega_{ab}\, \frac{dz^b}{d\tau} = \frac{\partial \mathcal{H}}{\partial z^a}, \tag{4}$$

where the Lagrange matrix $\boldsymbol{\omega}$ (with components $\omega_{ab} \equiv \partial_a \Gamma_b - \partial_b \Gamma_a$) is associated with the differential two-form

$$\omega = \mathsf{d}\Gamma = \mathsf{d}p_\mu \wedge \mathsf{d}x^\mu + \frac{e}{2c} \mathsf{F}_{\mu\nu}\, \mathsf{d}x^\mu \wedge \mathsf{d}x^\nu \equiv \frac{1}{2}\, \omega_{ab}\, \mathsf{d}z^a \wedge \mathsf{d}z^b. \tag{5}$$

The extended Poisson bracket $\{\ ,\ \}$ is obtained from the extended phase-space Lagrangian (2) by inverting the Lagrange matrix $\boldsymbol{\omega}$ to obtain the Poisson matrix $\mathbf{J} \equiv \boldsymbol{\omega}^{-1}$. From the Poisson-matrix components $J^{ab} \equiv \{z^a,\, z^b\}$, we obtain the extended Poisson bracket

$$\{F,\, G\} = \left(\frac{\partial F}{\partial x^\mu} \frac{\partial G}{\partial p_\mu} - \frac{\partial F}{\partial p_\mu} \frac{\partial G}{\partial x^\mu} \right) + \frac{e}{c}\, \mathsf{F}_{\mu\nu}\, \frac{\partial F}{\partial p_\mu} \frac{\partial G}{\partial p_\nu} \equiv \frac{\partial F}{\partial z^a}\, J^{ab}(\mathsf{z})\, \frac{\partial G}{\partial z^b}, \tag{6}$$

defined in terms of two arbitrary functions F and G. Hamilton's equations in extended phase space are thus expressed as

$$\frac{dz^a}{d\tau} = \{z^a,\, \mathcal{H}\} = J^{ab}(\mathsf{z})\, \frac{\partial \mathcal{H}(\mathsf{z})}{\partial z^b}, \tag{7}$$

which includes

$$\frac{dx^\mu}{dt} = \frac{\partial \mathcal{H}}{\partial p_\mu} \equiv v^\mu = \left(c,\ \mathbf{v} = \frac{\mathbf{p}}{m} \right), \tag{8}$$

and

$$\frac{dp_\mu}{dt} = -\frac{\partial \mathcal{H}}{\partial x^\mu} + \frac{e}{c}\, \mathsf{F}_{\mu\nu}\, \frac{\partial \mathcal{H}}{\partial p_\nu} \equiv \frac{e}{c}\, \mathsf{F}_{\mu\nu}\, v^\nu, \tag{9}$$

when $dt/d\tau = 1$ is substituted.

Next, the extended Vlasov equation, which describes the time evolution of the particle distribution on extended phase space, is expressed in terms of the extended Hamiltonian (1) and extended Poisson bracket (6) as

$$0 = \frac{d\mathcal{F}}{d\tau} = \frac{dz^a}{d\tau} \frac{\partial \mathcal{F}(\mathsf{z})}{\partial z^a} \equiv \{\mathcal{F},\, \mathcal{H}\}, \tag{10}$$

where, in order to satisfy the physical constraint (1), the extended Vlasov distribution is defined as

$$\mathcal{F}(\mathsf{z}) \equiv c\, \delta(w - |\mathbf{p}|^2/2m)\, f(\mathbf{x}, \mathbf{p}, t), \tag{11}$$

and $f(\mathbf{x}, \mathbf{p}, t)$ denotes the time-dependent Vlasov distribution on regular phase space. By integrating the extended Vlasov equation (10) over the energy coordinate w (and using $d\tau = dt$), we obtain the regular Vlasov equation

$$0 = \frac{df}{dt} \equiv \frac{\partial f}{\partial t} + \frac{d\mathbf{x}}{dt} \cdot \frac{\partial f}{\partial \mathbf{x}} + \frac{d\mathbf{p}}{dt} \cdot \frac{\partial f}{\partial \mathbf{p}}. \tag{12}$$

Lastly, the extended Vlasov equation (10) is coupled to Maxwell's equations

$$\nabla\cdot\mathbf{E} \;=\; 4\pi\,\rho, \tag{13}$$

$$\nabla\times\mathbf{B} - \frac{1}{c}\frac{\partial\mathbf{E}}{\partial t} \;=\; \frac{4\pi}{c}\,\mathbf{J}, \tag{14}$$

where $\mathbf{E} \equiv -\nabla\Phi - c^{-1}\partial\mathbf{A}/\partial t$ and $\mathbf{B} \equiv \nabla\times\mathbf{A}$ satisfy the constraints $\nabla\cdot\mathbf{B} = 0$ and $\nabla\times\mathbf{E} + c^{-1}\,\partial_t\mathbf{B} = 0$, and the charge-current densities

$$\left(\begin{array}{c}\rho\\ \mathbf{J}\end{array}\right) = \sum e\int d^4p\,\mathcal{F}\left(\begin{array}{c}1\\ \mathbf{v}\end{array}\right) = \sum e\int d^3p\,f\left(\begin{array}{c}1\\ \mathbf{v}\end{array}\right) \tag{15}$$

are defined in terms of the extended Vlasov distribution (11), with $d^4p = c^{-1}dw\,d^3p$.

2.2. *Vlasov-Maxwell variational principle*

The Vlasov-Maxwell equations (10) and (13)-(14) can be derived from the Vlasov-Maxwell action functional [19]

$$\mathcal{A}[\mathcal{F}, A_\mu] \;=\; -\sum\int d^8z\,\mathcal{F}(\mathsf{z})\,\mathcal{H}(\mathsf{z}) \;+\; \int\frac{d^4x}{16\,\pi}\,\mathsf{F}:\mathsf{F} \;\equiv\; \int d^4x\,\mathcal{L}(x). \tag{16}$$

The variation of the Lagrangian density

$$\mathcal{L}(x) \;\equiv\; \frac{1}{16\pi}\,\mathsf{F}(x):\mathsf{F}(x) \;-\; \sum\int d^4p\,\mathcal{F}(x,p)\,\mathcal{H}(p) \tag{17}$$

yields (after rearranging terms)

$$\delta\mathcal{L} \;\equiv\; \delta A_\mu\left(\frac{1}{4\pi}\frac{\partial\mathsf{F}^{\nu\mu}}{\partial x^\nu} + \sum\frac{e}{c}\int d^4p\,\mathcal{F}\,\frac{\partial\mathcal{H}}{\partial p_\mu}\right) - \sum\int d^4p\;\mathcal{S}\left\{\mathcal{F},\,\mathcal{H}\right\} + \frac{\partial\Lambda^\nu}{\partial x^\nu}. \tag{18}$$

To obtain this expression, we used the constraint $\mathsf{dF} \equiv 0$ for the electromagnetic field so that the field variation must be of the form $\delta\mathsf{F} \equiv \mathsf{d}\delta\mathsf{A}$. The constrained (Eulerian) variation for $\mathcal{F}$, on the other hand, is defined as

$$\delta\mathcal{F} \;\equiv\; \Delta\mathcal{F} \;-\; \delta z^a\,\partial_a\mathcal{F} \;=\; -\,\delta z^a\,\partial_a\mathcal{F}, \tag{19}$$

where the Lagrangian variation $\Delta\mathcal{F} \equiv 0$ (by definition, since the Vlasov distribution is constant along a *Lagrangian* orbit in phase space). The virtual displacement in extended phase-space is

$$\delta z^a \;\equiv\; -\left(\{\mathcal{S},\,z^a\} \;+\; \frac{e}{c}\,\delta A_\mu\;\{x^\mu, z^a\}\right), \tag{20}$$

where $\mathcal{S}$ is the canonical generating scalar field for this phase-space displacement. Hence, by substituting the displacement (20) into Eq. (19), we obtain the Eulerian variation

$$\delta\mathcal{F} \;\equiv\; \{\mathcal{S},\,\mathcal{F}\} \;+\; \frac{e}{c}\,\delta A_\mu\,\frac{\partial\mathcal{F}}{\partial p_\mu}. \tag{21}$$

Lastly, the last term in Eq. (18) involves the Noether four-vector

$$\Lambda^\nu \;\equiv\; \frac{1}{4\pi}\,\delta A_\mu\mathsf{F}^{\mu\nu} \;+\; \sum\int d^4p\,\mathcal{S}\left(\mathcal{F}\,\frac{\partial\mathcal{H}}{\partial p_\nu}\right), \tag{22}$$

which does not contribute to the variational principle

$$\int\,\delta\mathcal{L}\;d^4x \;=\; 0 \tag{23}$$

because this term appears as a space-time divergence in Eq. (18).

2.2.1. Vlasov-Maxwell Equations. We obtain the extended Vlasov equation (10) by simply requiring stationarity in Eq. (23) with respect to $\mathcal{S}$. Stationarity with respect to δA_ν, on the other hand, yields the Maxwell equations

$$\frac{1}{4\pi}\frac{\partial \mathsf{F}^{\nu\mu}}{\partial x^\nu} = -\sum \frac{e}{c}\int d^4p\,\frac{\partial \mathcal{H}}{\partial p_\mu}\,\mathcal{F} = -\sum \frac{e}{c}\int d^3p\; v^\mu\, f. \tag{24}$$

Decomposition in terms of components of F yields Eqs. (13)-(14).

2.2.2. Energy-Momentum Conservation Law. An important advantage of a variational formulation is that one can use Noether's Theorem [25] to derive explicit conservation laws. The conservation laws for the Vlasov-Maxwell equations (10) and (24) are derived from the Noether equation $\delta\mathcal{L} = \partial_\nu \Lambda^\nu$. By substituting explicit expressions for the variations $(\mathcal{S}, \delta A_\mu, \delta\mathcal{L})$, we can obtain the conservation laws of energy-momentum, angular momentum, and wave action. The latter conservation law plays a crucial role in the analysis of linear wave conversion in plasmas (see paper by Tracy and Brizard in this volume).

We now present the derivation of the energy-momentum conservation law associated with invariance of the Lagrangian density (17) with respect to space-time translation $x^\mu \to x^\mu + \delta x^\mu$. Under this translation, the variations $(\mathcal{S}, \delta A_\mu, \delta\mathcal{L})$ are

$$\left.\begin{array}{rcl}
\mathcal{S} & \equiv & (p_\mu + eA_\mu/c)\,\delta x^\mu \\
\delta A_\mu & \equiv & \mathsf{F}_{\mu\nu}\,\delta x^\nu - \partial_\mu\,(A_\nu\,\delta x^\nu) \\
\delta\mathcal{L} & \equiv & -\,\partial_\mu(\delta x^\mu\,\mathcal{L})
\end{array}\right\}, \tag{25}$$

where $\mathcal{S}$ generates the virtual displacement $\delta x^\mu = \{x^\mu,\, \mathcal{S}\}$ as required. After substituting the variations (25) in the Noether equation (22), and using the Maxwell equations (24), we obtain the energy-momentum conservation law $\partial_\mu\, \mathsf{T}^{\mu\nu} \equiv 0$, where the stress-energy tensor

$$\mathsf{T}^{\mu\nu} \equiv \frac{1}{4\pi}\left[\frac{1}{4}g^{\mu\nu}\,(\mathsf{F}:\mathsf{F}) - (\mathsf{F}\cdot\mathsf{F})^{\mu\nu}\right] + \sum\int d^4p\left(\frac{\partial\mathcal{H}}{\partial p_\mu}\,p^\nu\right)\mathcal{F} \tag{26}$$

is the sum of the field contribution (involving the tensor F) and the particle Vlasov contribution (involving the Vlasov distribution $\mathcal{F}$).

3. Reduced Plasma Dynamics

The processes of dynamical reduction in single-particle plasma dynamics and plasma kinetic theory are associated with the extended near-identity phase-space transformation $\mathcal{T}_\epsilon : \mathsf{z} \to \widehat{\mathsf{z}} = \mathcal{T}_\epsilon \mathsf{z}$, and its inverse $\mathcal{T}_\epsilon^{-1} : \widehat{\mathsf{z}} \to \mathsf{z} = \mathcal{T}_\epsilon^{-1}\widehat{\mathsf{z}}$. These transformations are expressed as asymptotic expansions in powers of a small dimensionless ordering parameter ϵ representing either a space-time scale ordering or the strength of the perturbation (e.g., wave amplitude) as [3]

$$\left.\begin{array}{rcl}
\widehat{z}^a & = & z^a + \epsilon\, G_1^a + \epsilon^2\left(G_2^a + \frac{1}{2}\mathsf{G}_1\cdot\mathsf{d}G_1^a\right) + \cdots \\
z^a & = & \widehat{z}^a - \epsilon\, G_1^a - \epsilon^2\left(G_2^a - \frac{1}{2}\mathsf{G}_1\cdot\mathsf{d}G_1^a\right) + \cdots
\end{array}\right\}. \tag{27}$$

Here, the nth-order vector field $\mathsf{G}_n\cdot\mathsf{d} = G_n^b\,\partial/\partial z^b$ is used to eliminate the fast space-time scales at order ϵ^n. Note that in standard applications of Lie-transform perturbation theory, the time coordinate is unchanged (i.e., $\widehat{t}\equiv t$) so that $G_n^t \equiv 0$ at all orders.

3.1. Reduced Push-forward and Pull-back Operators

The near-identity transformation (27) induces a transformation of the extended Vlasov equation (10) as follows. First, the push-forward operator T_ϵ^{-1} associated with the inverse phase-space transformation $\mathcal{T}_\epsilon^{-1}$, defined in Eq. (27), leads to the scalar-invariance relation

$$\widehat{\mathcal{F}}(\widehat{\mathsf{z}}) \;=\; \mathcal{F}(\mathsf{z}) \;=\; \mathcal{F}\left(\mathcal{T}_\epsilon^{-1}\widehat{\mathsf{z}}\right) \;\equiv\; \mathsf{T}_\epsilon^{-1}\mathcal{F}(\widehat{\mathsf{z}}) \tag{28}$$

between the particle Vlasov distribution $\mathcal{F}$ on the extended particle phase space and the reduced Vlasov distribution $\widehat{\mathcal{F}}$ on the extended reduced phase space.

Next, we construct the reduced extended Vlasov operator $d_\epsilon/d\tau$, defined in terms of the push-forward operator T_ϵ^{-1} (and its inverse, the pull-back operator T_ϵ) acting on an arbitrary function G as

$$\frac{d_\epsilon G}{d\tau} \;\equiv\; \mathsf{T}_\epsilon^{-1}\left[\frac{d}{d\tau}(\mathsf{T}_\epsilon\, G)\right] \;\equiv\; \{G,\,\widehat{\mathcal{H}}\}_\epsilon.$$

In the last expression, $d_\epsilon/d\tau$ is expressed in terms of the extended reduced Hamiltonian $\widehat{\mathcal{H}}$ and the extended reduced Poisson bracket

$$\left\{\widehat{F},\,\widehat{G}\right\}_\epsilon \;\equiv\; \mathsf{T}_\epsilon^{-1}\left(\left\{\mathsf{T}_\epsilon\widehat{F},\,\mathsf{T}_\epsilon\widehat{G}\right\}\right). \tag{29}$$

The reduced Poisson bracket (29) can also be constructed from the reduced phase-space Lagrangian $\widehat{\boldsymbol{\omega}}_\epsilon = \mathsf{d}\widehat{\Gamma}_\epsilon = \mathsf{d}\left(\mathsf{T}_\epsilon^{-1}\Gamma + \mathsf{d}\mathcal{S}\right) = \mathsf{T}_\epsilon^{-1}(\mathsf{d}\Gamma) = \mathsf{T}_\epsilon^{-1}\boldsymbol{\omega}$, so that $\widehat{J}_\epsilon^{ab} = (\widehat{\boldsymbol{\omega}}_\epsilon{}^{-1})^{ab} \equiv \{\widehat{z}^a,\,\widehat{z}^b\}_\epsilon$, where we used the fact that $\mathsf{d}^2\mathcal{S} = \partial^2_{ab}\mathcal{S}\,\mathsf{d}z^a\wedge\mathsf{d}z^b \equiv 0$ (i.e., $\nabla\times\nabla\mathcal{S}\equiv 0$ in three dimensions) and the exterior derivative d commutes with the operators T_ϵ and T_ϵ^{-1}.

Using the reduced extended Vlasov operator $d_\epsilon/d\tau$ induced by the near-identity phase-space transformation (27), the extended reduced Hamilton's equations are expressed as

$$\frac{d_\epsilon\widehat{z}^a}{d\tau} \;\equiv\; \left\{\widehat{z}^a,\,\widehat{\mathcal{H}}\right\}_\epsilon, \tag{30}$$

where the extended reduced Hamiltonian

$$\widehat{\mathcal{H}} \;\equiv\; \mathsf{T}_\epsilon^{-1}\mathcal{H} \;\equiv\; \widehat{H} - \widehat{w} \tag{31}$$

is defined as the push-forward of the extended Hamiltonian $\mathcal{H}\equiv H - w$, with $\widehat{H}\equiv\mathsf{T}_\epsilon^{-1}H - \partial\mathcal{S}/\partial t$. For many practical applications of reduced plasma dynamics, however, the new energy-momentum coordinates $\widehat{p}_\mu = (-\widehat{w}/c,\widehat{\mathbf{p}}_\mu)$ are canonical coordinates and, therefore, the reduced Poisson bracket $\{\;,\;\}_\epsilon \equiv \{\;,\;\}_{\rm c}$ is the canonical bracket $\{\widehat{F},\,\widehat{G}\}_{\rm c} = (\partial\widehat{F}/\partial\widehat{x}^\mu)\,(\partial\widehat{G}/\partial\widehat{p}_\mu) - (\partial\widehat{F}/\partial\widehat{p}_\mu)\,(\partial\widehat{G}/\partial\widehat{x}^\mu)$, which we now use in the remainder of this section.

Lastly, the near-identity phase-space transformation (27) also introduces the reduced-displacement vector

$$\boldsymbol{\rho}_\epsilon \;\equiv\; \mathsf{T}_\epsilon^{-1}\mathbf{x} \;-\; \widehat{\mathbf{x}}, \tag{32}$$

defined as the difference between the push-forward $\mathsf{T}_\epsilon^{-1}\mathbf{x}$ of the particle position $\mathbf{x}$ and the reduced position $\widehat{\mathbf{x}}$. The reduced displacement (32) is expressed as

$$\boldsymbol{\rho}_\epsilon \;=\; -\,\epsilon\,G_1^{\mathbf{x}} \;-\; \epsilon^2\left(G_2^{\mathbf{x}} - \frac{1}{2}\,\mathsf{G}_1\cdot\mathsf{d}G_1^{\mathbf{x}}\right) + \cdots \tag{33}$$

in terms of the generating vector fields $(\mathsf{G}_1, \mathsf{G}_2, \cdots)$ associated with the near-identity phase-space transformation (27). This reduced displacement plays in important role in the expression of polarization and magnetization effects in reduced plasma dynamics (see Sec. 3.4).

Journal of Physics: Conference Series **169** (2009) 012003 doi:10.1088/1742-6596/169/1/012003

3.2. Reduced Vlasov-Maxwell Equations

The extended reduced Vlasov equation for the extended reduced Vlasov distribution $\widehat{\mathcal{F}}$ is expressed as

$$0 = \left\{\widehat{\mathcal{F}},\, \widehat{\mathcal{H}}\right\}_c = \frac{d_\epsilon \widehat{\mathcal{F}}}{d\tau} = \frac{d_\epsilon \widehat{z}^a}{d\tau}\,\frac{\partial \widehat{\mathcal{F}}}{\partial \widehat{z}^a}. \tag{34}$$

Here, the reduced Vlasov distribution is defined as the push-forward of the particle Vlasov distribution

$$\widehat{\mathcal{F}}(\widehat{\mathbf{z}}) \equiv c\,\delta[\widehat{w} - \widehat{H}(\widehat{\mathbf{x}}, \widehat{\mathbf{p}}, t)]\, \widehat{F}(\widehat{\mathbf{x}}, \widehat{\mathbf{p}}, t), \tag{35}$$

with the reduced extended Hamiltonian (31) satisfying the physical constraint $\widehat{\mathcal{H}} \equiv 0$. When integrated over the reduced energy coordinate $\widehat{w}$, Eq. (34) becomes the regular reduced Vlasov equation

$$0 = \frac{d_\epsilon \widehat{F}}{dt} \equiv \frac{\partial \widehat{F}}{\partial t} + \frac{d_\epsilon \widehat{\mathbf{x}}}{dt}\cdot\frac{\partial \widehat{F}}{\partial \widehat{\mathbf{x}}} + \frac{d_\epsilon \widehat{\mathbf{p}}}{dt}\cdot\frac{\partial \widehat{F}}{\partial \widehat{\mathbf{p}}}, \tag{36}$$

where we used $dt/d\tau = 1$ to eliminate τ.

The dynamical reduction associated with the phase-space transformation (27) introduces polarization and magnetization effects into the Maxwell equations, which transforms the microscopic Maxwell's equations (13)-(14) into the macroscopic (reduced) Maxwell's equations [24, 26]

$$\nabla\cdot\mathbf{D}(x) = 4\pi\,\widehat{\rho}(x), \tag{37}$$

$$\nabla\times\mathbf{H}(x) - \frac{1}{c}\frac{\partial \mathbf{D}}{\partial t}(x) = \frac{4\pi}{c}\,\widehat{\mathbf{J}}(x). \tag{38}$$

Here, the terms "microscopic" and "macroscopic" are used to emphasize the fact that, while the sources $(\rho, \mathbf{J})$ of the microscopic Maxwell equations (13)-(14) are expressed in terms of moments of the particle Vlasov distribution $\mathcal{F}$, the sources $(\widehat{\rho}, \widehat{\mathbf{J}})$ of the macroscopic Maxwell equations (37)-(38) are expressed in terms of moments of the reduced Vlasov distribution $\widehat{\mathcal{F}}$. In Eqs. (37)-(38), the microscopic electric and magnetic fields $\mathbf{E}$ and $\mathbf{B}$ are replaced by the macroscopic fields [27]

$$\left.\begin{aligned} \mathbf{D}(x) &= \mathbf{E}(x) + 4\pi\,\mathbf{P}(x) \\ \mathbf{H}(x) &= \mathbf{B}(x) - 4\pi\,\mathbf{M}(x) \end{aligned}\right\}, \tag{39}$$

where $\mathbf{P}$ and $\mathbf{M}$ are the polarization and magnetization associated with the dynamical reduction (27). In the present paper, these polarization and magnetization are derived either by variational method (based on the existence of a reduced Lagrangian density) or by push-forward (Lie-transform) method (which establishes the connection between fluid moments in particle phase space and fluid moments in reduced phase space).

By comparing the microscopic and macroscopic (reduced) Maxwell's equations, we note that the dynamical reduction associated with the phase-space transformation (27) has introduced the following expressions for the charge and current densities:

$$\rho(x) \equiv \widehat{\rho}(x) - \nabla\cdot\mathbf{P}(x), \tag{40}$$

$$\mathbf{J}(x) \equiv \widehat{\mathbf{J}}(x) + \frac{\partial \mathbf{P}}{\partial t}(x) + c\,\nabla\times\mathbf{M}(x), \tag{41}$$

where $\rho_{\rm pol} \equiv -\nabla\cdot\mathbf{P}$ denotes the polarization density, $\mathbf{J}_{\rm pol} \equiv \partial\mathbf{P}/\partial t$ denotes the polarization current, and $\mathbf{J}_{\rm mag} \equiv c\,\nabla\times\mathbf{M}$ denotes the magnetization current. Note that the reduced charge-current densities $(\widehat{\rho}, \widehat{\mathbf{J}})$ satisfy the reduced charge conservation law

$$\frac{\partial\widehat{\rho}}{\partial t} + \nabla\cdot\widehat{\mathbf{J}} = \frac{\partial\rho}{\partial t} + \frac{\partial}{\partial t}\Big(\nabla\cdot\mathbf{P}\Big) + \nabla\cdot\mathbf{J} - \nabla\cdot\left(\frac{\partial\mathbf{P}}{\partial t} + c\,\nabla\times\mathbf{M}\right)$$

$$= \frac{\partial \rho}{\partial t} + \nabla \cdot \mathbf{J} \equiv 0. \tag{42}$$

The macroscopic (reduced) Maxwell's equations (37)-(38) can also be written in terms of the microscopic fields $(\mathbf{E}, \mathbf{B})$ as

$$\nabla \cdot \mathbf{E} = 4\pi \left(\widehat{\rho} - \nabla \cdot \mathbf{P}\right), \tag{43}$$

$$\nabla \times \mathbf{B} - \frac{1}{c}\frac{\partial \mathbf{E}}{\partial t} = \frac{4\pi}{c}\left(\widehat{\mathbf{J}} + \frac{\partial \mathbf{P}}{\partial t} + c\,\nabla \times \mathbf{M}\right), \tag{44}$$

where the polarization charge density and polarization-magnetization currents appear explicitly. It is this microscopic form that is most often useful for practical applications of reduced plasma dynamics [20].

3.3. Reduced Vlasov-Maxwell Variational Principle

We now show that the reduced Vlasov-Maxwell equations (34) and (43)-(44) can be derived from the reduced variational principle $\int d^4x\, \delta\widehat{\mathcal{L}} = 0$, where the reduced Lagrangian density is

$$\widehat{\mathcal{L}}(x) \equiv \frac{1}{16\pi}\mathsf{F}(x) : \mathsf{F}(x) - \sum \int d^4\widehat{p}\, \widehat{\mathcal{F}}(x, \widehat{p})\, \widehat{\mathcal{H}}(x, \widehat{p}; A, \mathsf{F}). \tag{45}$$

Note that, as a result of the dynamical reduction of the Vlasov equation, the reduced Hamiltonian appearing in Eq. (45) is expressed in terms of canonical energy-momentum coordinates as

$$\widehat{\mathcal{H}}(x, \widehat{p}; A, \mathsf{F}) \equiv \left(\frac{1}{2m}|\widehat{\mathbf{p}} - \frac{e}{c}\mathbf{A}(x)|^2 + e\,\Phi(x) - \widehat{w}\right) + \Psi_\epsilon\left(\widehat{\mathbf{p}} - \frac{e}{c}\mathbf{A}(x); \mathsf{F}(x)\right), \tag{46}$$

where the reduced *ponderomotive* potential Ψ_ϵ depends explicitly on the field tensor $\mathsf{F}_{\mu\nu}(x)$. From these field dependences $(\partial\widehat{\mathcal{H}}/\partial A_\mu, \partial\widehat{\mathcal{H}}/\partial \mathsf{F}_{\mu\nu})$, we define the reduced four-current density

$$\widehat{J}^\mu = (c\widehat{\rho}, \widehat{\mathbf{J}}) = -c \sum \int d^4\widehat{p}\, \widehat{\mathcal{F}}\, \frac{\partial\widehat{\mathcal{H}}}{\partial A_\mu} \equiv \sum e \int d^4\widehat{p}\, \widehat{\mathcal{F}}\, \frac{d_\epsilon \widehat{x}^\mu}{dt}, \tag{47}$$

where

$$\frac{d_\epsilon \widehat{x}^\mu}{dt} \equiv \frac{\partial\widehat{\mathcal{H}}}{\partial \widehat{p}_\mu} = \left(c,\; \frac{1}{m}(\widehat{\mathbf{p}} - \frac{e}{c}\mathbf{A}) + \frac{\partial\Psi_\epsilon}{\partial\widehat{\mathbf{p}}}\right). \tag{48}$$

The reduced antisymmetric polarization-magnetization tensor [17], on the other hand, is defined as

$$\mathsf{K}^{\mu\nu} \equiv -2\sum \int d^4\widehat{p}\, \widehat{\mathcal{F}}\, \frac{\partial\Psi_\epsilon}{\partial \mathsf{F}_{\mu\nu}}, \tag{49}$$

where the polarization and magnetization $\mathsf{K}^{0i} = P^i$ and $\mathsf{K}^{ij} = \epsilon^{ijk} M_k$ are defined as

$$(\mathbf{P}, \mathbf{M}) \equiv -\sum \int d^4\widehat{p}\, \widehat{\mathcal{F}} \left(\frac{\partial\Psi_\epsilon}{\partial\mathbf{E}}, \frac{\partial\Psi_\epsilon}{\partial\mathbf{B}}\right). \tag{50}$$

The reduced ponderomotive potential Ψ_ϵ therefore contains all the polarization and magnetization effects associated with the phase-space transformation (27).

3.3.1. Reduced Vlasov-Maxwell Variational Principle. By following the same steps outlined in Sec. 2.2, we begin our variational formulation of the reduced Vlasov-Maxwell equations with an expression for the variation of the Lagrangian density

$$\delta\widehat{\mathcal{L}} \;=\; \partial_{\mu}\widehat{\Lambda}^{\mu} \;-\; \sum\int d^{4}\widehat{p}\,\widehat{\mathcal{S}}\,\{\widehat{\mathcal{F}},\,\widehat{\mathcal{H}}\} \;+\; \frac{\delta A_{\nu}}{4\pi}\left[\,\partial_{\mu}(\mathsf{F}^{\mu\nu} \;+\; 4\pi\,\mathsf{K}^{\mu\nu}) \;+\; \frac{4\pi}{c}\,\widehat{J}^{\nu}\,\right], \tag{51}$$

where the reduced Noether four-vector

$$\widehat{\Lambda}^{\mu} \;\equiv\; \sum\int d^{4}\widehat{p}\;\widehat{\mathcal{S}}\,\widehat{\mathcal{F}}\;\frac{\partial\widehat{\mathcal{H}}}{\partial\widehat{p}_{\mu}} \;+\; \frac{\delta A_{\nu}}{4\pi}(\mathsf{F}^{\nu\mu} + 4\pi\,\mathsf{K}^{\nu\mu}) \tag{52}$$

does not contribute to the variational principle $\int d^{4}x\,\delta\widehat{\mathcal{L}} = 0$. As a result of the reduced variational principle, where the variations $\widehat{\mathcal{S}}$ and δA_{ν} are arbitrary, we obtain the reduced Vlasov equation (34) and the reduced (*macroscopic*) Maxwell equations

$$\frac{\partial}{\partial x^{\mu}}\,(\mathsf{F}^{\mu\nu} \;+\; 4\pi\,\mathsf{K}^{\mu\nu}) \;=\; -\,\frac{4\pi}{c}\,\widehat{J}^{\nu}, \tag{53}$$

from which we recover the reduced Maxwell equations (43) and (44). In addition, we find that the reduced charge conservation law (42) follows immediately from the reduced Maxwell equations (53), since the tensors $\mathsf{F}^{\mu\nu}$ and $\mathsf{K}^{\mu\nu}$ are antisymmetric.

3.3.2. Reduced Energy-Momentum Conservation Law. We now derive the energy-momentum conservation law from the reduced Noether equation $\delta\widehat{\mathcal{L}} \equiv \partial_{\mu}\widehat{\Lambda}^{\mu}$ associated with space-time translations $x^{\mu} \rightarrow x^{\mu} + \delta x^{\mu}$. The corresponding Eulerian variations $(\widehat{\mathcal{S}}, \delta A_{\nu}, \delta\widehat{\mathcal{L}})$ are

$$\left.\begin{array}{rcl}
\widehat{\mathcal{S}} & = & \widehat{p}_{\mu}\,\delta x^{\mu} \\
\delta A_{\mu} & = & \mathsf{F}_{\mu\nu}\,\delta x^{\nu} - \partial_{\mu}(A_{\nu}\,\delta x^{\nu}) \\
\delta\widehat{\mathcal{L}} & = & -\,\partial_{\mu}(\delta x^{\mu}\,\widehat{\mathcal{L}})
\end{array}\right\}. \tag{54}$$

After some cancellations introduced through the reduced Maxwell equations (53), we obtain the reduced energy-momentum conservation law $\partial_{\mu}\mathsf{T}^{\mu\nu} \equiv 0$, where the reduced stress-energy tensor $\mathsf{T}^{\mu\nu}$ is defined as

$$\mathsf{T}^{\mu\nu} \;\equiv\; \frac{1}{4\pi}\left[\,\frac{g^{\mu\nu}}{4}\,\mathsf{F}:\mathsf{F} \;-\; (\mathsf{F}^{\mu\sigma} + 4\pi\,\mathsf{K}^{\mu\sigma})\;\mathsf{F}_{\sigma}^{\;\;\nu}\,\right] \;+\; \sum\int d^{4}\widehat{p}\;\frac{\partial\widehat{\mathcal{H}}}{\partial\widehat{p}_{\mu}}\;\left(\widehat{p}^{\nu} - \frac{e}{c}\,A^{\nu}\right)\,\widehat{\mathcal{F}} \tag{55}$$

which naturally includes the polarization and magnetization (50). An explicit proof of the reduced energy-momentum conservation law is given in Ref. [24]. Note that, because of polarization and magnetization effects ($\Psi_{\epsilon} \neq 0$), the reduced stress-energy tensor (55) does not appear to be symmetric (while it may be so when evaluated explicitly). The physical meaning of this unsymmetrical Minkowski form [27] for the reduced stress-energy tensor has a rich history in physics (briefly discussed by Boghosian [17] and recently reviewed in Refs. [28, 29]) but its further discussion is beyond the scope of this paper.

3.4. Push-forward Definitions of Polarization and Magnetization

A complementary method for deriving the polarization and magnetization (50) is provided by the push-forward method [24]. From the scalar-invariance relation (28), we construct the push-forward relation of fluid moments of an arbitrary function χ on particle phase space as follows.

First, we transform the momentum average $[\chi]$ of an arbitrary phase-space function χ:

$$\begin{aligned} n\,[\chi] &\equiv \int d^3p\, f\,\chi \;=\; \int d^8z\,\mathcal{F}\,\delta^4(x-r)\,\chi \\ &= \int d^8\widehat{z}\,\widehat{\mathcal{F}}\;\mathsf{T}_\epsilon^{-1}\left(\delta^4(x-r)\,\chi\right) \\ &\equiv \int d^6\widehat{z}\,\widehat{F}\,\delta^3(\widehat{\mathbf{x}}+\boldsymbol{\rho}_\epsilon-\mathbf{r})\,\mathsf{T}_\epsilon^{-1}\chi, \end{aligned} \tag{56}$$

where the extended Vlasov distribution (35) is used (n denotes the particle fluid density) and energy-time integrations are carried out. The reduced displacement (33) has components that are dependent on the fast space-time scales and components that are independent. Next, by expanding the right side of Eq. (56) in powers of $\boldsymbol{\rho}_\epsilon$ and integrating by parts, we obtain the push-forward relation

$$n\,[\chi] \equiv \widehat{n}\,\left[\mathsf{T}_\epsilon^{-1}\chi\right]^\wedge - \nabla\cdot\left(\widehat{n}\,\left[\boldsymbol{\rho}_\epsilon\,\mathsf{T}_\epsilon^{-1}\chi\right]^\wedge - \nabla\cdot\frac{\widehat{n}}{2}\,\left[\boldsymbol{\rho}_\epsilon\boldsymbol{\rho}_\epsilon\,\mathsf{T}_\epsilon^{-1}\chi\right]^\wedge + \cdots\right), \tag{57}$$

where $[\cdots]^\wedge$ denotes an average with respect to the reduced Vlasov distribution $\widehat{F}$ in reduced phase space ($\widehat{n}$ denotes the reduced fluid density). In Eq. (57), the terms in the divergence on the right side include the dipolar contribution (linear in $\boldsymbol{\rho}_\epsilon$) and the quadrupolar contribution (quadratic in $\boldsymbol{\rho}_\epsilon$). We now apply the push-forward relation (57) to derive dipolar expressions for the polarization and magnetization (where the quadrupolar and higher-order multipole contributions are ignored).

First, we consider the case $\chi = e\;\;(= \mathsf{T}_\epsilon^{-1}\chi)$ so that the push-forward relation (57) yields the expression for the charge density

$$\rho = \widehat{\rho} - \nabla\cdot\left(\sum e\,\widehat{n}\,[\boldsymbol{\rho}_\epsilon]^\wedge + \cdots\right), \tag{58}$$

where $\widehat{\rho} \equiv \sum e\,\widehat{n}$ defines the reduced charge density and higher-order multipole terms (e.g., quadrupole term that is quadratic in $\boldsymbol{\rho}_\epsilon$) are not displayed. By comparing Eq. (40) with the relation (58) between the particle charge density ρ and the reduced charge density $\widehat{\rho}$, we find the expression for the polarization (dipole-moment density)

$$\mathbf{P} \equiv \sum\left(\widehat{n}\,[\boldsymbol{\pi}_\epsilon]^\wedge\right) = \sum \int d^4\widehat{p}\,\boldsymbol{\pi}_\epsilon\,\widehat{\mathcal{F}}, \tag{59}$$

where the reduced electric-dipole moment

$$\boldsymbol{\pi}_\epsilon \equiv e\,\boldsymbol{\rho}_\epsilon \tag{60}$$

is associated with the charge displacement induced by the phase-space transformation (27). Note that the fast-time average $\langle\boldsymbol{\pi}_\epsilon\rangle = e\,\langle\boldsymbol{\rho}_\epsilon\rangle$ of the reduced electric-dipole moment (60) may not vanish and thus may provide important polarization effects [5, 20].

Next, we consider the case $\chi = e\mathbf{v} \equiv e\,d\mathbf{x}/dt$, where the Lagrangian representation for the particle's velocity $\mathbf{v} = d\mathbf{x}/dt$ is used. To obtain an expression for the push-forward $\mathsf{T}_\epsilon^{-1}(e\,d\mathbf{x}/dt)$, we use the reduced Vlasov operator d_ϵ/dt, defined in Eq. (34), to obtain

$$e\,\mathsf{T}_\epsilon^{-1}\mathbf{v} = e\,\mathsf{T}_\epsilon^{-1}\frac{d\mathbf{x}}{dt} = e\left(\mathsf{T}_\epsilon^{-1}\frac{d}{dt}\mathsf{T}_\epsilon\right)\mathsf{T}_\epsilon^{-1}\mathbf{x} \equiv e\,\frac{d_\epsilon}{dt}\left(\mathsf{T}_\epsilon^{-1}\mathbf{x}\right) = e\left(\frac{d_\epsilon\widehat{\mathbf{x}}}{dt} + \frac{d_\epsilon\boldsymbol{\rho}_\epsilon}{dt}\right), \tag{61}$$

where Eq. (32) is used to obtain the last expression. The reduced (e.g., guiding-center) velocity $d_\epsilon\widehat{\mathbf{x}}/dt$ is independent of the fast space-time scales, while the reduced displacement velocity

$d_\epsilon \boldsymbol{\rho}_\epsilon/dt$ exhibits both fast and slow space-time scales. Note that the fast-time-average particle *polarization* velocity $d_\epsilon\langle\boldsymbol{\rho}_\epsilon\rangle/dt$ may not vanish [30] and thus may represent additional reduced dynamical effects (e.g., the standard polarization drift in guiding-center theory [6]) not included in the reduced velocity $d_\epsilon\widehat{\mathbf{x}}/dt$.

We now replace the reduced moment $[d_\epsilon\boldsymbol{\rho}_\epsilon/dt]^\wedge$ in the first term of Eq. (57) by taking the time derivative of the polarization (59) and using the reduced Vlasov equation (36) to obtain [24]

$$\begin{aligned}
\frac{\partial \mathbf{P}}{\partial t} &= \sum e \int d^4\widehat{p}\left(\frac{\partial \boldsymbol{\rho}_\epsilon}{\partial t}\,\widehat{\mathcal{F}} + \boldsymbol{\rho}_\epsilon\,\frac{\partial \widehat{\mathcal{F}}}{\partial t}\right) \\
&= \sum e\,\widehat{n}\left[\frac{d_\epsilon\boldsymbol{\rho}_\epsilon}{dt}\right]^\wedge - \nabla\cdot\left(\sum e\,\widehat{n}\left[\frac{d_\epsilon\widehat{\mathbf{x}}}{dt}\,\boldsymbol{\rho}_\epsilon\right]^\wedge\right),
\end{aligned} \tag{62}$$

where integration by parts of the reduced Vlasov equation

$$\frac{\partial \widehat{\mathcal{F}}}{\partial t} = -\widehat{\nabla}\widehat{\mathcal{F}}\cdot\frac{d_\epsilon\widehat{\mathbf{x}}}{dt} - \frac{\partial\widehat{\mathcal{F}}}{\partial\widehat{\mathbf{p}}}\cdot\frac{d_\epsilon\widehat{\mathbf{p}}}{dt} \equiv -\widehat{\nabla}\cdot\left(\widehat{\mathcal{F}}\,\frac{d_\epsilon\widehat{\mathbf{x}}}{dt}\right) - \frac{\partial}{\partial\widehat{\mathbf{p}}}\cdot\left(\widehat{\mathcal{F}}\,\frac{d_\epsilon\widehat{\mathbf{p}}}{dt}\right)$$

is carried out to obtain the last expression. Hence, the push-forward relation (57) for $\chi = e\,d\mathbf{x}/dt$ yields the expression for the current density [24]

$$\begin{aligned}
\mathbf{J} &= \widehat{\mathbf{J}} + \frac{\partial\mathbf{P}}{\partial t} + \nabla\times\left(\sum e\,\widehat{n}\left[\boldsymbol{\rho}_\epsilon\times\frac{d_\epsilon\widehat{\mathbf{x}}}{dt}\right]^\wedge\right) + \nabla\times\left(\sum\frac{e}{2}\,\widehat{n}\left[\boldsymbol{\rho}_\epsilon\times\frac{d_\epsilon\boldsymbol{\rho}_\epsilon}{dt}\right]^\wedge\right) \\
&\equiv \widehat{\mathbf{J}} + \frac{\partial\mathbf{P}}{\partial t} + c\,\nabla\times\mathbf{M}
\end{aligned} \tag{63}$$

where $\widehat{\mathbf{J}} \equiv \sum e\,\widehat{n}\,[d_\epsilon\widehat{\mathbf{x}}/dt]^\wedge$ denotes the reduced current density, the second term represents the polarization current $\mathbf{J}_{\rm pol} \equiv \partial\mathbf{P}/\partial t$, and the third term represents the magnetization current $\mathbf{J}_{\rm mag} \equiv c\,\nabla\times\mathbf{M}$. By comparing Eq. (41) with the relation (63), we obtain the magnetization

$$\mathbf{M} = \sum\frac{e}{c}\int d^4\widehat{p}\;\boldsymbol{\rho}_\epsilon\times\left(\frac{1}{2}\,\frac{d_\epsilon\boldsymbol{\rho}_\epsilon}{dt} + \frac{d_\epsilon\widehat{\mathbf{x}}}{dt}\right)\widehat{\mathcal{F}} \equiv \sum\int d^4\widehat{p}\,\left(\boldsymbol{\mu}_\epsilon + \frac{\boldsymbol{\pi}_\epsilon}{c}\times\frac{d_\epsilon\widehat{\mathbf{x}}}{dt}\right)\widehat{\mathcal{F}}, \tag{64}$$

which is represented as the sum (for each particle species) of an intrinsic magnetic-dipole contribution

$$\boldsymbol{\mu}_\epsilon \equiv \frac{e}{2c}\left(\boldsymbol{\rho}_\epsilon\times\frac{d_\epsilon\boldsymbol{\rho}_\epsilon}{dt}\right), \tag{65}$$

and a moving electric-dipole [27] contribution ($\boldsymbol{\pi}_\epsilon\times d_\epsilon\widehat{\mathbf{x}}/dt$). We note that, since the relation $n' \equiv \gamma\,n$ between the laboratory density n' and the rest-frame density n involves the relativistic factor $\gamma = 1/\sqrt{1-|\mathbf{v}|^2/c^2}$, the standard moving-dipole magnetization contribution [31] to the polarization appearing in the charge density (58) comes from the last term in the relativistic c^{-2} correction

$$\begin{aligned}
e\,\boldsymbol{\rho}_\epsilon\,\mathsf{T}_\epsilon^{-1}(\gamma-1) &= \frac{e}{2c^2}\,\boldsymbol{\rho}_\epsilon\,\left|\mathsf{T}_\epsilon^{-1}\mathbf{v}\right|^2 + \cdots \\
&= \frac{e}{2c^2}\,\boldsymbol{\rho}_\epsilon\left(\left|\frac{d_\epsilon\widehat{\mathbf{x}}}{dt}\right|^2 + \left|\frac{d_\epsilon\boldsymbol{\rho}_\epsilon}{dt}\right|^2\right) + \frac{e}{c^2}\,\frac{d_\epsilon\widehat{\mathbf{x}}}{dt}\cdot\left(\frac{d_\epsilon\boldsymbol{\rho}_\epsilon}{dt}\,\boldsymbol{\rho}_\epsilon\right),
\end{aligned} \tag{66}$$

where the first term yields a small relativistic correction to the electric dipole moment. We return to this relativistic magnetization effect in Sec. 4.5 [see Eq. (131)].

A simple example of the magnetization (65) is provided by guiding-center dynamics [5, 6], where $\boldsymbol{\rho}$ denotes the gyroangle-dependent gyroradius vector (with $d_\epsilon \boldsymbol{\rho}_\epsilon/dt = \Omega\, \partial\boldsymbol{\rho}/\partial\theta$ to lowest order, where Ω denotes the gyrofrequency) and the guiding-center intrinsic magnetization $\boldsymbol{\mu}_{\rm gc} \equiv -\mu\,\widehat{\sf b}$ is defined in terms of the magnetic moment $\mu \equiv m|\mathbf{v}_\perp|^2/2B$ (an adiabatic invariant for guiding-center motion) and $\widehat{\sf b} \equiv \mathbf{B}/B$. The guiding-center electric dipole moment [5], on the other hand, is expressed as $\boldsymbol{\pi}_{\rm gc} \equiv e\,(\widehat{\sf b}/\Omega) \times \mathbf{v}_{\rm gc}$, where $\mathbf{v}_{\rm gc}$ denotes the guiding-center drift velocity (excluding the $E \times B$ velocity).

3.5. Polarization and Magnetization

We now summarize the main results of this Section. The polarization and magnetization

$$\begin{aligned} \mathbf{P} &= -\sum \widehat{n}\, \frac{\partial[\Psi_\epsilon]^\wedge}{\partial \mathbf{E}} = \sum \widehat{n}\, [\boldsymbol{\pi}_\epsilon]^\wedge, \\ \mathbf{M} &= -\sum \widehat{n}\, \frac{\partial[\Psi_\epsilon]^\wedge}{\partial \mathbf{B}} = \sum \widehat{n}\, \left[\boldsymbol{\mu}_\epsilon + \frac{\boldsymbol{\pi}_\epsilon}{c} \times \frac{d_\epsilon \widehat{\mathbf{x}}}{dt}\right]^\wedge, \end{aligned}$$

can either be derived by variational method from the reduced potential Ψ_ϵ or by push-forward method from the reduced displacement $\boldsymbol{\rho}_\epsilon \equiv {\sf T}_\epsilon^{-1}\mathbf{x} - \widehat{\mathbf{x}}$. By combining these expressions, we obtain the relations between the reduced dipole moments $(\boldsymbol{\pi}_\epsilon, \boldsymbol{\mu}_\epsilon)$ and the partial derivatives $(\partial\Psi_\epsilon/\partial\mathbf{E}, \partial\Psi_\epsilon/\partial\mathbf{B})$ of the reduced Hamiltonian:

$$\left.\begin{aligned} -\,\partial\Psi_\epsilon/\partial\mathbf{E} &\equiv \boldsymbol{\pi}_\epsilon \\ -\,\partial\Psi_\epsilon/\partial\mathbf{B} &\equiv \boldsymbol{\mu}_\epsilon + (\boldsymbol{\pi}_\epsilon/c) \times d_\epsilon\widehat{\mathbf{x}}/dt \end{aligned}\right\}. \tag{67}$$

These relations emphasize the complementarity of the reduced variational and push-forward methods.

4. Oscillation-center dynamics in weakly-magnetized plasmas

The problem of the low-frequency oscillation-center Hamiltonian dynamics of charged particles in high-frequency, short-wavelength electromagnetic waves is the paradigm for applications of Lie-transform perturbation methods in plasma physics [10-13]. Here, the fast wave space-time scales are removed asymptotically from the Hamiltonian dynamics by a time-dependent phase-space noncanonical transformation $z^a = (ct, \mathbf{x}; w/c, \mathbf{p}) \rightarrow \overline{z}^a = (ct, \overline{\mathbf{x}}; \overline{w}/c, \overline{\mathbf{p}})$, where $\overline{\mathbf{x}}$ denotes the oscillation-center position, and $(\overline{w}, \overline{\mathbf{p}} \equiv m\,\overline{\mathbf{v}})$ denote the oscillation-center's kinetic energy-momentum. Note that, although the phase-space transformation is time-dependent, the time coordinate t remains unchanged by the transformation (i.e., the particle and oscillation-center times are identical).

First, we decompose the electromagnetic potentials $(\Phi, \mathbf{A})$ in powers of the wave amplitude (represented by the ordering parameter ϵ):

$$\begin{pmatrix} \Phi \\ \mathbf{A} \end{pmatrix} = \sum_{n=0}^{\infty} \epsilon^n \begin{pmatrix} \Phi_n \\ \mathbf{A}_n \end{pmatrix}, \tag{68}$$

where the lowest-order background plasma is represented by the zeroth-order fields $(\Phi_0, \mathbf{A}_0)$, while the primary wave fields $(\mathbf{E}_1, \mathbf{B}_1)$ are represented by the first-order potentials $(\Phi_1, \mathbf{A}_1)$ and the second-order (back-reaction) electromagnetic fields $(\mathbf{E}_2, \mathbf{B}_2)$ are represented by the second-order potentials $(\Phi_2, \mathbf{A}_2)$. These second-order fields, which represents the plasma response to the external first-order fields, contain contributions that are weakly space-time dependent compared to the rapid wave space-time scales, and contributions with rapid *second-harmonic* wave space-time scales.

The extended-phase-space Hamiltonian dynamics of charged particles in the electromagnetic fields (68) is expressed in terms of the extended (unperturbed) Hamiltonian

$$\mathcal{H} \;=\; \frac{1}{2m}\,|\mathbf{p}|^2 \;-\; w \;\equiv\; \mathcal{H}_0, \tag{69}$$

and the extended (perturbed) phase-space Lagrangian

$$\begin{aligned}\Gamma \;&=\; \left[\mathbf{p} + \frac{e}{c}\left(\mathbf{A}_0 + \epsilon\,\mathbf{A}_1 + \epsilon^2\,\mathbf{A}_2 + \cdots\right)\right]\cdot\mathsf{d}\mathbf{x} \;-\; \left[w + e\left(\Phi_0 + \epsilon\,\Phi_1 + \epsilon^2\,\Phi_2 + \cdots\right)\right]\mathsf{d}t \\ &\equiv\; \Gamma_0 \,+\, \epsilon\,\Gamma_1 \,+\, \epsilon^2\,\Gamma_2 \,+\, \cdots. \end{aligned} \tag{70}$$

Here, the unperturbed Poisson bracket $\{\;,\;\}_0$ is obtained from Γ_0 as

$$\{F,\,G\}_0 \;\equiv\; \left(\frac{\partial F}{\partial x^\mu}\,\frac{\partial G}{\partial p_\mu} \;-\; \frac{\partial F}{\partial p_\mu}\,\frac{\partial G}{\partial x^\mu}\right) + \frac{e}{c}\,\mathsf{F}_{(0)\mu\nu}\;\frac{\partial F}{\partial p_\mu}\,\frac{\partial G}{\partial p_\nu}. \tag{71}$$

4.1. Eikonal Representation

We now introduce the eikonal representation for the electromagnetic fields (68). The previous work by Cary and Kaufman [11] considered the case of an unmagnetized background plasma (with $\Phi_0 = 0 = \mathbf{A}_0$) and the case of a strongly magnetized plasma (with $\Phi_0 = 0$). Here, we consider the intermediate case of a weakly-magnetized background plasma described as follows.

First, we assume that the zeroth-order (background) fields $A^\mu_{(0)} \equiv (\Phi_0,\,\mathbf{A}_0)$ are weakly space-time dependent:

$$A^\mu_{(0)} \;\equiv\; A^\mu_{(0)}(\epsilon_0 x), \tag{72}$$

where $\epsilon_0 \ll 1$ denotes the eikonal ordering parameter (warning: ϵ_0 is not the permitivity of free space), so that the background electric and magnetic fields

$$\mathsf{F}_{(0)\mu\nu} \;\equiv\; \epsilon_0\,\mathsf{F}_{(0)\mu\nu}(\epsilon_0 x) \tag{73}$$

are ordered small; the weak-background-field (WBF) ordering (73) is consistent with a weakly-magnetized quasi-static background plasma. In particular, according to this ordering, the wave frequency ω is considered much larger than the gyrofrequency $\Omega_0 = eB_0/mc$ of a charged particle moving in this weak magnetic field while the wavelength $\lambda \equiv 2\pi\,|\mathbf{k}|^{-1}$ is considered much smaller than the particle's gyroradius ρ_0. Because of the mass dependence associated with this ordering, the weak-background-field (WBF) ordering (73) is especially appropriate for ions (because of their large masses), while the ordering might be invalid for electrons (because of their small masses).

Next, the first-order wave fields are decomposed in terms of the eikonal representation:

$$\begin{pmatrix}\Phi_1\\ \mathbf{A}_1\end{pmatrix} \;\equiv\; \begin{pmatrix}\widetilde{\Phi}_1\\ \widetilde{\mathbf{A}}_1\end{pmatrix} e^{i\,\Theta/\epsilon_0} \;+\; \begin{pmatrix}\widetilde{\Phi}_1^*\\ \widetilde{\mathbf{A}}_1^*\end{pmatrix} e^{-\,i\,\Theta/\epsilon_0}, \tag{74}$$

where the wave eikonal-amplitudes (denoted by a tilde) are weakly-varying space-time functions and derivatives of the eikonal phase $\Theta(\epsilon_0\mathbf{r},\,\epsilon_0 t)$:

$$\epsilon_0^{-1}\,\frac{\partial\Theta}{\partial x^\mu}(\epsilon_0 x) \;\equiv\; k_\mu(\epsilon_0 x) \;=\; (-\omega/c,\;\mathbf{k}) \tag{75}$$

define the weakly-varying wavevector $\mathbf{k}$ and wave frequency ω. Note that the first-order electromagnetic field has the eikonal-amplitude

$$\widetilde{\mathsf{F}}_{(1)\mu\nu} \;\equiv\; i\left(k_\mu\,\widetilde{A}_{(1)\nu} \;-\; k_\nu\,\widetilde{A}_{(1)\mu}\right). \tag{76}$$

Lastly, the eikonal representation of the second-order fields is

$$\begin{pmatrix} \Phi_2 \\ \mathbf{A}_2 \end{pmatrix} \equiv \begin{pmatrix} \overline{\Phi}_2 \\ \overline{\mathbf{A}}_2 \end{pmatrix} + \begin{pmatrix} \widetilde{\Phi}_2 \\ \widetilde{\mathbf{A}}_2 \end{pmatrix} e^{2i\,\Theta/\epsilon_0} + \begin{pmatrix} \widetilde{\Phi}_2^* \\ \widetilde{\mathbf{A}}_2^* \end{pmatrix} e^{-2i\,\Theta/\epsilon_0}, \tag{77}$$

where the eikonal-averaged weakly-varying back-reaction fields $(\overline{\Phi}_2, \overline{\mathbf{A}}_2)$ are generated by the second-order *ponderomotive* effects (as discussed below) while second-harmonic terms $(\widetilde{\Phi}_2, \widetilde{\mathbf{A}}_2)$ are, henceforth, ignored (i.e., an eikonal-averaged expression involving second-harmonic terms appears at the fourth-order in wave amplitude, which falls outside the scope of our work).

4.2. Oscillation-center Transformation and Ponderomotive Hamiltonian

As a result of the perturbation expansion (68) in the electromagnetic potentials $(\Phi, \mathbf{A})$, we see that the Hamiltonian (69) exhibits explicit dependence on the eikonal phase Θ and, thus, the particle dynamics exhibits fast space-time scales in addition to the slow space-time scales associated with the background plasma.

Lie-transform Hamiltonian perturbation theory [3] is concerned with the derivation of a new extended oscillation-center Hamiltonian

$$\overline{\mathcal{H}} \equiv \mathsf{T}_\epsilon^{-1}\mathcal{H} = \left(\overline{H}_0 + \epsilon\,\overline{H}_1 + \epsilon^2\,\overline{H}_2 + \cdots\right) - \overline{w}, \tag{78}$$

where the first-order and second-order oscillation-center Hamiltonians are

$$\overline{H}_1 = H_1 - \mathsf{G}_1\cdot\mathsf{d}\mathcal{H}_0 \equiv -\,\mathsf{G}_1\cdot\mathsf{d}\mathcal{H}_0, \tag{79}$$

$$\overline{H}_2 = H_2 - \mathsf{G}_2\cdot\mathsf{d}\mathcal{H}_0 - \frac{1}{2}\,\mathsf{G}_1\cdot\mathsf{d}(H_1+\overline{H}_1) \equiv -\,\mathsf{G}_2\cdot\mathsf{d}\mathcal{H}_0 - \frac{1}{2}\,\mathsf{G}_1\cdot\mathsf{d}\overline{H}_1. \tag{80}$$

The last expressions in Eqs. (79)-(80) use the fact that the Hamiltonian perturbation terms H_n $(n \geq 1)$ are identically zero in the Hamiltonian (69). The generating vector fields $(\mathsf{G}_1, \mathsf{G}_2, \cdots)$, on the other hand, are determined from the requirement that the oscillation-center phase-space Lagrangian

$$\overline{\Gamma} \equiv \mathsf{T}_\epsilon^{-1}\Gamma + \mathsf{d}\mathcal{S} \equiv \overline{\Gamma}_0, \tag{81}$$

retains its unperturbed form. From this requirement, we obtain

$$G_1^a = \{S_1,\, z^a\}_0 + \frac{e}{c}\,A_{(1)\mu}\,\{x^\mu,\, z^a\}_0, \tag{82}$$

$$G_2^a = \{S_2,\, z^a\}_0 + \frac{e}{c}\,A_{(2)\mu}\,\{x^\mu,\, z^a\}_0 - \frac{e}{2c}\,\{S_1,\, x^\mu\}_0\,\mathsf{F}_{(1)\mu\nu}\,\{x^\nu,\, z^a\}_0, \tag{83}$$

where the scalar fields $(S_1, S_2, \cdots)$ are determined from Eqs. (79)-(80) by requiring that the oscillation-center Hamiltonian (78) be eikonal-phase independent.

4.2.1. First-order analysis. From Eqs. (79) and (82), we obtain the first-order Hamiltonian

$$\overline{H}_1 = -\,\{S_1,\, \mathcal{H}_0\}_0 - \frac{e}{c}\,A_{(1)\mu}\,\{x^\mu,\, \mathcal{H}_0\}_0 = -\,\frac{dS_1}{dt} + e\left(\Phi_1 - \frac{\mathbf{v}}{c}\cdot\mathbf{A}_1\right) \equiv 0, \tag{84}$$

where the first-order oscillation-center Hamiltonian $\overline{H}_1$ must vanish since $(\Phi_1, \mathbf{A}_1, S_1)$ are all explicitly eikonal-phase dependent. By inserting the eikonal representation (74) into Eq. (84), we obtain the eikonal equation for $\widetilde{S}_1$:

$$-i\,\omega'\,\widetilde{S}_1 + \epsilon_0\,e\left(\mathbf{E}_0 + \frac{\mathbf{v}}{c}\times\mathbf{B}_0\right)\cdot\frac{\partial\widetilde{S}_1}{\partial\mathbf{p}} = e\left(\widetilde{\Phi}_1 - \frac{\mathbf{v}}{c}\cdot\widetilde{\mathbf{A}}_1\right) \tag{85}$$

where $\omega' = \omega - \mathbf{k}\cdot\mathbf{v}$ is the Doppler-shifted wave frequency, and the WBF eikonal ordering (73) has been used to obtain the second term on the left side. Next, by substituting the WBF-eikonal expansion

$$\widetilde{S}_1 \;=\; \widetilde{S}_{10} \;+\; \epsilon_0\,\widetilde{S}_{11} \;+\; \cdots \tag{86}$$

into Eq. (85), we obtain the lowest-order eikonal amplitude

$$\widetilde{S}_{10} \;=\; \frac{ie}{\omega'}\left(\widetilde{\Phi}_1 \;-\; \frac{\mathbf{v}}{c}\cdot\widetilde{\mathbf{A}}_1\right), \tag{87}$$

and the first-order WBF-eikonal correction

$$\widetilde{S}_{11} \;=\; -\,\frac{ie}{\omega'}\left(\mathbf{E}_0 + \frac{\mathbf{v}}{c}\times\mathbf{B}_0\right)\cdot\widetilde{\boldsymbol{\xi}}, \tag{88}$$

where the first-order spatial displacement $\widetilde{\boldsymbol{\xi}}$ is defined in terms of the lowest-order eikonal amplitude [12]

$$\widetilde{\boldsymbol{\xi}} \equiv \frac{\partial\widetilde{S}_{10}}{\partial\mathbf{p}} \;=\; -\,\frac{e}{m\omega'^2}\left(\widetilde{\mathbf{E}}_1 + \frac{\mathbf{v}}{c}\times\widetilde{\mathbf{B}}_1\right). \tag{89}$$

Note that the first-order spatial-displacement eikonal amplitude $\widetilde{\boldsymbol{\xi}}$ also satisfies the eikonal relation $i\,m\omega'\,\widetilde{\boldsymbol{\xi}} = i\mathbf{k}\,\widetilde{S}_{10} + e\widetilde{\mathbf{A}}_1/c$, which can be expressed as the first-order field equation

$$-\,m\,\frac{d_0\boldsymbol{\xi}}{dt} \;=\; \nabla S_{10} \;+\; \frac{e}{c}\,\mathbf{A}_1, \tag{90}$$

where $d_0/dt \equiv \partial/\partial t + \mathbf{v}\cdot\nabla$ and $\omega' \equiv -\,\epsilon_0^{-1}d_0\Theta/dt$.

The phase-space transformation (27) can thus be expressed explicitly as

$$\left.\begin{array}{rcl}
\overline{\mathbf{x}} & = & \mathbf{x} - \epsilon\,\partial S_{10}/\partial\mathbf{p} + \cdots \;=\; \mathbf{x} \;-\; \epsilon\,\boldsymbol{\xi} \;+\; \cdots \\
\overline{\mathbf{p}} & = & \mathbf{p} + \epsilon\,(\nabla S_{10} + e\mathbf{A}_1/c) + \cdots \;=\; \mathbf{p} \;-\; \epsilon\,m\,d_0\boldsymbol{\xi}/dt \;+\; \cdots \\
\overline{w} & = & w - \epsilon\,(\partial S_{10}/\partial t - e\Phi_1) + \cdots \;=\; w - \epsilon\,m\mathbf{v}\cdot d_0\boldsymbol{\xi}/dt \;+\; \cdots
\end{array}\right\}, \tag{91}$$

where $d_0S_{10}/dt = e\,(\Phi_1 - \mathbf{A}_1\cdot\mathbf{v}/c)$. Hence, to first order in ϵ, the oscillation-center momentum coordinate is

$$\overline{\mathbf{p}} \;=\; m\left(\mathbf{v} - \epsilon\,\frac{d_0\boldsymbol{\xi}}{dt}\right) \;+\; \cdots \;\equiv\; m\,\frac{d\overline{\mathbf{x}}}{dt},$$

and the oscillation-center energy coordinate is

$$\overline{w} \;=\; \frac{m}{2}\,|\mathbf{v}|^2 \;-\; \epsilon\,m\mathbf{v}\cdot\frac{d_0\boldsymbol{\xi}}{dt} \;+\; \cdots \;\equiv\; \frac{1}{2m}\,|\overline{\mathbf{p}}|^2,$$

which satisfies the unperturbed constraint $\overline{\mathcal{H}}_0 \;=\; |\overline{\mathbf{p}}|^2/2m - \overline{w} \equiv 0$ for the oscillation-center kinetic energy-momentum coordinates.

4.2.2. Second-order analysis. Next, from Eqs. (80), (83), and (84), the expression for the second-order Hamiltonian is

$$\begin{aligned}
\overline{H}_2 \;&\equiv\; -\,\{S_2,\,\mathcal{H}_0\}_0 \;-\; \frac{e}{c}\,A_{(2)\mu}\,\{x^\mu,\,\mathcal{H}_0\}_0 \;-\; \frac{e}{2c}\,\frac{\partial S_1}{\partial p_\mu}\,\mathsf{F}_{(1)\mu\nu}\,\{x^\nu,\,\mathcal{H}_0\}_0 \\
&=\; -\,\frac{dS_2}{dt} \;+\; e\left(\Phi_2 - \frac{\mathbf{v}}{c}\cdot\mathbf{A}_2\right) \;-\; \frac{e}{2}\left(\mathbf{E}_1 + \frac{\mathbf{v}}{c}\times\mathbf{B}_1\right)\cdot\frac{\partial S_1}{\partial\mathbf{p}},
\end{aligned} \tag{92}$$

where the right side has both eikonal-dependent and eikonal-independent terms. To derive the eikonal-independent terms that define the second-order oscillation-center Hamiltonian $\overline{H}_2$, we perform an eikonal-phase average of the right side of Eq. (92) to obtain

$$\overline{H}_2 \;=\; e\left(\overline{\Phi}_2 \;-\; \frac{\mathbf{v}}{c}\cdot\overline{\mathbf{A}}_2\right) \;-\; \frac{e}{2}\left\langle\left(\mathbf{E}_1 + \frac{\mathbf{v}}{c}\times\mathbf{B}_1\right)\cdot\frac{\partial S_1}{\partial\mathbf{p}}\right\rangle, \tag{93}$$

where the contributions of the second-harmonic eikonal-amplitudes $(\widetilde{\Phi}_2, \widetilde{\mathbf{A}}_2, \widetilde{S}_2 \equiv S_2)$ vanish.

Using the definition (89), the lowest-order contribution from the second term in Eq. (93) is

$$-\,\frac{e}{2}\left\langle\left(\mathbf{E}_1 + \frac{\mathbf{v}}{c}\times\mathbf{B}_1\right)\cdot\frac{\partial S_{10}}{\partial\mathbf{p}}\right\rangle \;=\; m\omega'^2\,|\widetilde{\boldsymbol{\xi}}|^2, \tag{94}$$

which determines the standard ponderomotive Hamiltonian [11]. Using the definition (88), on the other hand, the first-order WBF correction to the ponderomotive Hamiltonian is

$$\begin{aligned} -\,\frac{e}{2}\left\langle\left(\mathbf{E}_1 + \frac{\mathbf{v}}{c}\times\mathbf{B}_1\right)\cdot\frac{\partial S_{11}}{\partial\mathbf{p}}\right\rangle \;&=\; \frac{\partial}{\partial\mathbf{p}}\cdot\mathrm{Re}\left(m\omega'^2\widetilde{\boldsymbol{\xi}}\,\widetilde{S}_{11}^*\right) \;=\; \frac{\partial}{\partial\mathbf{p}}\cdot\left[\,\frac{i}{2}\,m\omega'\,\mathbf{F}_0\times\left(\widetilde{\boldsymbol{\xi}}\times\widetilde{\boldsymbol{\xi}}^*\right)\right] \\ &\equiv\; -\left(\mathbf{E}_0 + \frac{\mathbf{v}}{c}\times\mathbf{B}_0\right)\cdot\overline{\boldsymbol{\pi}}_2 \;-\; \mathbf{B}_0\cdot\overline{\boldsymbol{\mu}}_2, \end{aligned} \tag{95}$$

where we introduced the definitions for the ponderomotive electric dipole moment

$$\overline{\boldsymbol{\pi}}_2 \;\equiv\; e\,\mathbf{k}\times\left(i\,\widetilde{\boldsymbol{\xi}}\times\widetilde{\boldsymbol{\xi}}^*\right), \tag{96}$$

and the ponderomotive magnetic dipole moment

$$\overline{\boldsymbol{\mu}}_2 \;\equiv\; \frac{e}{c}\,\omega'\,\left(i\,\widetilde{\boldsymbol{\xi}}\times\widetilde{\boldsymbol{\xi}}^*\right). \tag{97}$$

Note that the intrinsic ponderomotive magnetic-dipole moment $\overline{\boldsymbol{\mu}}_2$ is accompanied in Eq. (95) by the ponderomotive moving-electric-dipole moment $\overline{\boldsymbol{\pi}}_2\times\mathbf{v}/c$ [27]. As indicated above, however, the moving-magnetic-dipole contribution $(\mathbf{v}/c)\times\overline{\boldsymbol{\mu}}_2$ to the ponderomotive polarization appears as a relativistic correction, which is ignored in the present nonrelativistic treatment.

4.2.3. Ponderomotive Hamiltonian. By combining Eqs. (94)-(95), the second-order oscillation-center Hamiltonian is defined by the eikonal-averaged terms on the right side of Eq. (92):

$$\begin{aligned} \overline{H}_2 \;&=\; e\left(\overline{\Phi}_2 \;-\; \frac{\overline{\mathbf{v}}}{c}\cdot\overline{\mathbf{A}}_2\right) \;+\; m\omega'^2\,|\widetilde{\boldsymbol{\xi}}|^2 \;-\; \epsilon_0\left[\,\mathbf{E}_0\cdot\overline{\boldsymbol{\pi}}_2 \;+\; \mathbf{B}_0\cdot\left(\overline{\boldsymbol{\mu}}_2 \;+\; \overline{\boldsymbol{\pi}}_2\times\frac{\overline{\mathbf{v}}}{c}\right)\right] \\ &\equiv\; e\left(\overline{\Phi}_2 \;-\; \frac{\overline{\mathbf{v}}}{c}\cdot\overline{\mathbf{A}}_2\right) \;+\; \overline{\Psi}_2, \end{aligned} \tag{98}$$

where the reduced potential $\overline{\Psi}_2 = \overline{\Psi}_{20} + \epsilon_0\,\overline{\Psi}_{21}$ contains the ponderomotive potential $\overline{\Psi}_{20}$ [Eq. (94)] and its first-order (WBF) correction $\overline{\Psi}_{21}$ [Eq. (95)], which includes the ponderomotive polarization and magnetization effects (96)-(97).

Lastly, by using the ponderomotive dipole moments (96) and (97), we can immediately obtain the following expressions for the ponderomotive polarization and magnetization

$$\left(\overline{\mathbf{P}}_2,\,\overline{\mathbf{M}}_2\right) \;\equiv\; \sum\int d^3\overline{p}\left(\overline{\boldsymbol{\pi}}_2,\;\overline{\boldsymbol{\mu}}_2 \;+\; \overline{\boldsymbol{\pi}}_2\times\frac{\overline{\mathbf{v}}}{c}\right)\,\overline{F} \;=\; -\,\sum\int d^3\overline{p}\left(\frac{\partial\overline{\Psi}_{21}}{\partial\mathbf{E}_0},\,\frac{\partial\overline{\Psi}_{21}}{\partial\mathbf{B}_0}\right)\,\overline{F}, \tag{99}$$

which satisfy the relations (67) between the variational and push-forward methods. Note that the polarization and magnetization

$$\mathbf{P} \;=\; \epsilon\,\mathbf{P}_1 \;+\; \epsilon^2\,\mathbf{P}_2 \;+\;\cdots \quad\text{and}\quad \mathbf{M} \;=\; \epsilon\,\mathbf{M}_1 \;+\; \epsilon^2\,\mathbf{M}_2 \;+\;\cdots \tag{100}$$

have first-order contributions that vanish upon eikonal-phase averaging (i.e., $\langle\mathbf{P}_1\rangle = 0 = \langle\mathbf{M}_1\rangle$), while the eikonal-averaged contributions are second-order in wave amplitude: $\langle\mathbf{P}\rangle = \epsilon^2\,\overline{\mathbf{P}}_2$ and $\langle\mathbf{M}\rangle = \epsilon^2\,\overline{\mathbf{M}}_2$.

4.3. Oscillation-center Pull-back Operator

The relation between the particle Vlasov distribution f and the oscillation-center Vlasov distribution $\overline{F}$ is expressed in terms of the pull-back operator T_ϵ:

$$\begin{aligned} f &= \mathsf{T}_\epsilon \overline{F} \equiv \overline{F} + \epsilon\, \mathsf{G}_1 \cdot \mathsf{d}\overline{F} + \epsilon^2 \left[\mathsf{G}_2 \cdot \mathsf{d}\overline{F} + \frac{1}{2}\, \mathsf{G}_1 \cdot \mathsf{d}\left(\mathsf{G}_1 \cdot \mathsf{d}\overline{F} \right) \right] + \cdots \\ &\equiv f_0 + \epsilon\, f_1 + \epsilon^2\, f_2 + \cdots, \end{aligned} \tag{101}$$

where the vectors $(\mathsf{G}_1, \mathsf{G}_2, \cdots)$ are defined in Eqs. (82) and (83).

4.3.1. First-order pull-back operator. Since the oscillation-center Vlasov distribution $\overline{F}$ is independent of the fast wave space-time scales (i.e., it is independent of the eikonal phase Θ), the eikonal-phase averaged part $\langle f_1 \rangle$ of the first-order particle Vlasov distribution $f_1 \equiv \mathsf{G}_1 \cdot \mathsf{d}\overline{F}$ vanishes: $\langle f_1 \rangle \equiv 0$ (since $\langle \mathsf{G}_1 \rangle \equiv 0$). Its eikonal-phase dependent part $\widetilde{f}_1 = \widetilde{f}_{10} + \epsilon_0\, \widetilde{f}_{11} + \cdots$, on the other hand, is expressed in terms of the lowest-order contribution

$$\widetilde{f}_{10} = i\, m\omega'\, \widetilde{\boldsymbol{\xi}} \cdot \frac{\partial \overline{F}}{\partial \mathbf{p}} = -\,\frac{ie}{\omega'} \left(\widetilde{\mathbf{E}}_1 + \frac{\mathbf{v}}{c} \times \widetilde{\mathbf{B}}_1 \right) \cdot \frac{\partial \overline{F}}{\partial \mathbf{p}}. \tag{102}$$

The eikonal-amplitude of the first-order charge density

$$\sum e \int \widetilde{f}_{10}\, d^3p = \sum -i\,e \int d^3\overline{p}\, \frac{\partial}{\partial \overline{\mathbf{p}}} \cdot \left(m\omega' \widetilde{\boldsymbol{\xi}} \right) \overline{F} = -\,i\,\mathbf{k} \cdot \left(\int d^3\overline{p}\, \widetilde{\boldsymbol{\xi}}\, \overline{F} \right), \tag{103}$$

yields the eikonal-amplitude for the first-order polarization

$$\widetilde{\mathbf{P}}_1 \equiv \sum e \int \overline{F}\, \widetilde{\boldsymbol{\xi}}\, d^3\overline{p}. \tag{104}$$

The eikonal-amplitude of the first-order current density, on the other hand, is expressed as

$$\begin{aligned} \sum e \int \mathbf{v}\, \widetilde{f}_{10}\, d^3p &= \sum -i\,e \int d^3\overline{p}\, \frac{\partial}{\partial \overline{\mathbf{p}}} \cdot \left(m\omega' \widetilde{\boldsymbol{\xi}}\, \overline{\mathbf{v}} \right) \overline{F} \\ &= -\sum e \int d^3\overline{p} \left[\left(i\mathbf{k} \cdot \widetilde{\boldsymbol{\xi}} \right) \overline{\mathbf{v}} + i\,\omega'\, \widetilde{\boldsymbol{\xi}} \right] \overline{F} \\ &= \sum e \int d^3\overline{p} \left[-i\,\omega\, \widetilde{\boldsymbol{\xi}} + i\,\mathbf{k} \times \left(\widetilde{\boldsymbol{\xi}} \times \overline{\mathbf{v}} \right) \right] \overline{F}, \end{aligned} \tag{105}$$

where the first term in the last expression yields the eikonal-amplitude of the first-order polarization current, while the second term yields the eikonal-amplitude of the first-order magnetization

$$\widetilde{\mathbf{M}}_1 \equiv \sum e \int d^3\overline{p} \left(\widetilde{\boldsymbol{\xi}} \times \frac{\overline{\mathbf{v}}}{c} \right) \overline{F}. \tag{106}$$

Note that only the moving electric-dipole contribution appears in Eq. (106) since the intrinsic magnetization must be quadratic in $\boldsymbol{\xi}$.

4.3.2. Second-order pull-back operator. The second-order particle Vlasov distribution $f_2 \equiv \mathsf{G}_2 \cdot \mathsf{d}\overline{F} + \frac{1}{2}\, \mathsf{G}_1 \cdot \mathsf{d}(\mathsf{G}_1 \cdot \mathsf{d}\overline{F})$ has an eikonal-phase independent part expressed as

$$\langle f_{20} \rangle = \frac{\partial}{\partial \mathbf{p}} \cdot \left[\frac{e}{c} \left(\overline{\mathbf{A}}_2 + \mathrm{Re}(\widetilde{\mathbf{B}}_1 \times \widetilde{\boldsymbol{\xi}}^*) \right) \overline{F} + m^2 \omega'^2\, \mathrm{Re}\left(\widetilde{\boldsymbol{\xi}}\, \widetilde{\boldsymbol{\xi}}^* \right) \cdot \frac{\partial \overline{F}}{\partial \mathbf{p}} \right], \tag{107}$$

where only the lowest-order WBF terms are kept. Note that this second-order particle contribution has a vanishing zeroth velocity-moment

$$\int \langle f_{20}\rangle\, d^3p \;\equiv\; 0, \tag{108}$$

since Eq. (107) is expressed as a momentum-space divergence. However, the first velocity-moment

$$\begin{aligned}\int \mathbf{v}\,\langle f_{20}\rangle\, d^3p \;&=\; -\int d^3\overline{p}\left[\left(\frac{e\overline{\mathbf{A}}_2}{mc} + \frac{e}{mc}\,\mathrm{Re}\left(\widetilde{\mathbf{B}}_1\times\widetilde{\boldsymbol{\xi}}^*\right)\right)\overline{F} \;+\; m\omega'^{\,2}\,\mathrm{Re}(\widetilde{\boldsymbol{\xi}}\,\widetilde{\boldsymbol{\xi}}^*)\cdot\frac{\partial\overline{F}}{\partial\overline{\mathbf{p}}}\right] \\ &=\; \int d^3\overline{p}\left[-\,\frac{e}{mc}\,\overline{\mathbf{A}}_2 \;+\; \omega'\,\mathbf{k}\cdot\left(\widetilde{\boldsymbol{\xi}}\,\widetilde{\boldsymbol{\xi}}^* + \widetilde{\boldsymbol{\xi}}^*\widetilde{\boldsymbol{\xi}}\right)\right]\overline{F}\end{aligned} \tag{109}$$

is non-vanishing, where the following identities are used in obtaining the last expression: $\partial_{\mathbf{p}}(m\omega'\,\widetilde{\boldsymbol{\xi}}) \equiv \widetilde{\boldsymbol{\xi}}\,\mathbf{k}$ and

$$\frac{e}{c}\,\mathrm{Re}\left(\widetilde{\mathbf{B}}_1\times\widetilde{\boldsymbol{\xi}}^*\right) \;\equiv\; m\omega'\,\left[\,|\widetilde{\boldsymbol{\xi}}|^2\,\mathbf{k} \;-\; \mathbf{k}\cdot\mathrm{Re}(\widetilde{\boldsymbol{\xi}}^*\widetilde{\boldsymbol{\xi}})\,\right],$$

which follows from the relation $(e/c)\,\widetilde{\mathbf{B}}_1 \;=\; -\,m\omega'\;\mathbf{k}\times\widetilde{\boldsymbol{\xi}}$ obtained from Eq. (89). Using the lowest-order second-order oscillation-center Hamiltonian (98), we obtain the identity

$$\int \mathbf{v}\,\langle f_{20}\rangle\, d^3p \;\equiv\; \int\left(-\,\frac{e}{mc}\,\overline{\mathbf{A}}_2 \;+\; \frac{\partial\overline{\Psi}_{20}}{\partial\overline{\mathbf{p}}}\right)\overline{F}\;d^3\overline{p}. \tag{110}$$

Lastly, the second-order kinetic energy

$$\int d^3p\,\left(\frac{|\mathbf{p}|^2}{2m}\right)\,\langle f_{20}\rangle \;\equiv\; \int\left(\overline{\Psi}_{20} \;-\; \frac{e}{c}\,\overline{\mathbf{A}}_2\cdot\overline{\mathbf{v}}\right)\overline{F}\;d^3\overline{p} \tag{111}$$

is naturally expressed in terms of the oscillation-center distribution $\overline{F}$ and the (lowest-order) second-order oscillation-center Hamiltonian (98).

4.4. Ponderomotive Variational Principle

The variational formulation for the exact Vlasov-Maxwell equations was presented in Sec. 2. The covariance of the Vlasov part of the action functional has been used in Ref. [32] to derive a variational formulation of the nonlinear gyrokinetic Vlasov-Maxwell equations, which describe the turbulent evolution of low-frequency electromagnetic fluctuations in a magnetized plasma with arbitrary geometry [8].

4.4.1. Linear polarization and magnetization. The eikonal form of the first-order Maxwell equations can be expressed in terms of the eikonal-averaged second-order Lagrangian density

$$\begin{aligned}\mathcal{L}_{20} \;&=\; \frac{1}{4\pi}\left(|\widetilde{\mathbf{E}}_1|^2 \;-\; |\widetilde{\mathbf{B}}_1|^2\right) \;-\; \sum\int d^3\overline{p}\,\left[\,e\left(\overline{\Phi}_2 \;-\; \frac{\overline{\mathbf{v}}}{c}\cdot\overline{\mathbf{A}}_2\right) \;+\; \overline{\Psi}_{20}\,\right]\overline{F} \\ &\equiv\; \frac{1}{4\pi}\,\mathrm{Re}\left(\widetilde{\mathbf{E}}_1\cdot\widetilde{\mathbf{D}}_1^* \;-\; \widetilde{\mathbf{B}}_1\cdot\widetilde{\mathbf{H}}_1^*\right) \;-\; \sum e\int d^3\overline{p}\,\left(\overline{\Phi}_2 \;-\; \frac{\overline{\mathbf{v}}}{c}\cdot\overline{\mathbf{A}}_2\right)\overline{F},\end{aligned} \tag{112}$$

where the last expression is obtained by writing the ponderomotive potential $\overline{\Psi}_{20}$ in terms of the first-order polarization and magnetization (104) and (106) as

$$\sum\int d^3\overline{p}\,\overline{F}\,\overline{\Psi}_{20} \;\equiv\; -\,\mathrm{Re}\left(\widetilde{\mathbf{P}}_1^*\cdot\widetilde{\mathbf{E}}_1 \;+\; \widetilde{\mathbf{M}}_1^*\cdot\widetilde{\mathbf{B}}_1\right),$$

which displays the standard relation between the ponderomotive Hamiltonian and the oscillation-center polarization and magnetization [33, 34].

We define the first-order eikonal-amplitudes for the macroscopic electromagnetic fields [27]

$$\widetilde{\mathbf{D}}_1 \;\equiv\; 4\pi\,\frac{\partial\mathcal{L}_{20}}{\partial\widetilde{\mathbf{E}}_1^*} \;\equiv\; \widetilde{\mathbf{E}}_1 \;+\; 4\pi\,\widetilde{\mathbf{P}}_1, \tag{113}$$

$$\widetilde{\mathbf{H}}_1 \;\equiv\; -\,4\pi\,\frac{\partial\mathcal{L}_{20}}{\partial\widetilde{\mathbf{B}}_1^*} \;\equiv\; \widetilde{\mathbf{B}}_1 \;-\; 4\pi\,\widetilde{\mathbf{M}}_1, \tag{114}$$

in terms of the first-order eikonal-amplitudes for the polarization and magnetization (104) and (106). Hence, by using these definitions, the first-order Maxwell equations become

$$0 \;=\; \frac{\partial\mathcal{L}_{20}}{\partial\widetilde{\Phi}_1^*} \;\equiv\; \frac{\partial\widetilde{\mathbf{E}}_1^*}{\partial\widetilde{\Phi}_1^*}\cdot\frac{\partial\mathcal{L}_{20}}{\partial\widetilde{\mathbf{E}}_1^*} \;=\; i\,\mathbf{k}\cdot\frac{\widetilde{\mathbf{D}}_1}{4\pi}, \tag{115}$$

$$0 \;=\; \frac{\partial\mathcal{L}_{20}}{\partial\widetilde{\mathbf{A}}_1^*} \;\equiv\; \frac{\partial\widetilde{\mathbf{E}}_1^*}{\partial\widetilde{\mathbf{A}}_1^*}\cdot\frac{\partial\mathcal{L}_{20}}{\partial\widetilde{\mathbf{E}}_1^*} \;+\; \frac{\partial\widetilde{\mathbf{B}}_1^*}{\partial\widetilde{\mathbf{A}}_1^*}\cdot\frac{\partial\mathcal{L}_{20}}{\partial\widetilde{\mathbf{B}}_1^*} \;=\; -\,i\,\frac{\omega}{c}\,\frac{\widetilde{\mathbf{D}}_1}{4\pi} \;-\; i\,\mathbf{k}\times\frac{\widetilde{\mathbf{H}}_1}{4\pi}, \tag{116}$$

where the eikonal amplitudes for the first-order charge-current densities are

$$\begin{aligned}\begin{pmatrix}\widetilde{\rho}_1\\ \widetilde{\mathbf{J}}_1\end{pmatrix} &\equiv \sum e\int d^3p\,\begin{pmatrix}1\\ \mathbf{v}\end{pmatrix}\widetilde{f}_{10} = \sum e\int d^3\overline{p}\,\begin{pmatrix}-i\,\mathbf{k}\cdot\widetilde{\boldsymbol{\xi}}\\ -i\,\omega\,\widetilde{\boldsymbol{\xi}} \,+\, i\,\mathbf{k}\times\left(\widetilde{\boldsymbol{\xi}}\times\overline{\mathbf{v}}\right)\end{pmatrix}\overline{F}\\ &= \begin{pmatrix}-i\,\mathbf{k}\cdot\widetilde{\mathbf{P}}_1\\ -i\,\omega\,\widetilde{\mathbf{P}}_1 \,+\, i\,\mathbf{k}c\times\widetilde{\mathbf{M}}_1\end{pmatrix},\end{aligned} \tag{117}$$

which are consistent with Eqs. (104) and (106). Note that Eq. (116) is consistent with Eq. (115), i.e., the dot product of Eq. (116) with the wave vector **k** yields Eq. (115).

4.4.2. Nonlinear polarization and magnetization. The derivation of the lowest-order expressions for the eikonal-averaged second-order charge and current densities, and the eikonal corrections corresponding to the ponderomotive polarization and magnetization, proceeds with the introduction of the second-order eikonal-averaged Vlasov-Maxwell action functional

$$\begin{aligned}\mathcal{L}_2 \;=\;& \frac{1}{4\pi}\left[\mathrm{Re}\left(\widetilde{\mathbf{E}}_1\cdot\widetilde{\mathbf{D}}_1^* \,-\, \widetilde{\mathbf{B}}_1\cdot\widetilde{\mathbf{H}}_1^*\right) \,+\, \epsilon_0\left(\mathbf{E}_0\cdot\overline{\mathbf{D}}_2 \,-\, \mathbf{B}_0\cdot\overline{\mathbf{H}}_2\right)\right]\\ &-\sum e\int d^3\overline{p}\,\left(\overline{\Phi}_2 \,-\, \frac{\overline{\mathbf{v}}}{c}\cdot\overline{\mathbf{A}}_2\right)\overline{F}.\end{aligned} \tag{118}$$

We first note that, since the second-order eikonal-averaged Vlasov Lagrangian density (118) is independent of the background scalar potential Φ_0, we find

$$\overline{\rho}_{20} \;\equiv\; -\,\frac{\partial\mathcal{L}_2}{\partial\Phi_0} \;=\; 0, \tag{119}$$

and, thus, the second-order oscillation-center eikonal-averaged charge density must be generated by ponderomotive polarization effects [see Eqs. (122)-(123)] to first order (in ϵ_0) in eikonal analysis. This result is consistent with the fact that the zeroth-moment (108) of the second-order eikonal-averaged particle Vlasov distribution $\langle f_{20}\rangle$ vanishes.

The second-order ponderomotive (eikonal-averaged) polarization and magnetization are obtained from the action functional (118) as

$$\overline{\mathbf{P}}_2 \;=\; \epsilon_0^{-1}\frac{\partial\mathcal{L}_2}{\partial\mathbf{E}_0} - \frac{\overline{\mathbf{E}}_2}{4\pi} \;=\; \sum\int \overline{\boldsymbol{\pi}}_2\,\overline{F}\,d^3\overline{p}, \tag{120}$$

$$\overline{\mathbf{M}}_2 \;=\; \epsilon_0^{-1}\frac{\partial\mathcal{L}_2}{\partial\mathbf{B}_0} + \frac{\overline{\mathbf{B}}_2}{4\pi} \;=\; \sum\int \left(\overline{\boldsymbol{\mu}}_2 \,+\, \overline{\boldsymbol{\pi}}_2\times\frac{\overline{\mathbf{v}}}{c}\right)\,\overline{F}\,d^3\overline{p}, \tag{121}$$

where $\overline{\boldsymbol{\pi}}_2$ and $\overline{\boldsymbol{\mu}}_2$ are defined in Eqs. (96) and (97). The second-order ponderomotive charge and current densities are

$$\overline{\rho}_{21} \;=\; -\,\nabla\cdot\overline{\mathbf{P}}_2, \tag{122}$$

$$\overline{\mathbf{J}}_{21} \;=\; \frac{\partial\overline{\mathbf{P}}_2}{\partial t} \,+\, c\,\nabla\times\overline{\mathbf{M}}_2, \tag{123}$$

respectively.

Lastly, the second-order eikonal-averaged Maxwell equations can also be written as

$$\nabla\cdot\overline{\mathbf{D}}_2 \;=\; 0 \quad\text{and}\quad \nabla\times\overline{\mathbf{H}}_2 \,-\, \frac{1}{c}\frac{\partial\overline{\mathbf{D}}_2}{\partial t} \;=\; \frac{4\pi}{c}\,\overline{\mathbf{J}}_{20}, \tag{124}$$

which requires $\overline{\mathbf{J}}_{20}$ to be divergenceless [11], as is also required for the eikonal-averaged second-order charge and current densities $\overline{\rho}_2 = \overline{\rho}_{20} + \epsilon_0\,\overline{\rho}_{21}$ and $\overline{\mathbf{J}}_2 = \overline{\mathbf{J}}_{20} + \epsilon_0\,\overline{\mathbf{J}}_{21}$ to satisfy the eikonal-averaged second-order charge conservation law

$$\frac{\partial\overline{\rho}_2}{\partial t} \,+\, \nabla\cdot\overline{\mathbf{J}}_2 \;=\; 0. \tag{125}$$

Hence, the condition (119) immediately implies that $\nabla\cdot\overline{\mathbf{J}}_{20} \equiv 0$ [11], while the second-order ponderomotive charge and current densities (122)-(123) explicitly satisfy Eq. (125).

We have, thus, achieved the purpose of this Section, which was to show how the ponderomotive polarization and magnetization (99) can be self-consistently derived from a variational principle (118) for the oscillation-center Vlasov-Maxwell equations. We note that ponderomotive polarization and magnetization effects in magnetized plasmas have also been investigated in Refs. [14, 16] by using a Low-Lagrangian-type variational formulation.

4.5. Ponderomotive Push-forward Derivation

We now proceed with the complementary derivation of the ponderomotive polarization and magnetization by the ponderomotive push-forward method. By inserting the oscillation-center transformation generated by Eqs. (82) and (83) into the reduced displacement (33), we obtain the oscillation-center displacement

$$\boldsymbol{\rho}_\epsilon \;\equiv\; \epsilon\,\boldsymbol{\xi} \,+\, \frac{\epsilon^2}{2}\left(\boldsymbol{\xi}\cdot\nabla\boldsymbol{\xi} \,+\, m\,\frac{d\boldsymbol{\xi}}{dt}\cdot\frac{\partial\boldsymbol{\xi}}{\partial\mathbf{p}}\right) \,-\, \epsilon^2\,G_2^{\mathbf{x}} \,+\, \cdots, \tag{126}$$

where Eq. (90) is used and the first-order WBF corrections are omitted in what follows. While the second-order term $G_2^{\mathbf{x}} \equiv -\,\partial_{\mathbf{p}}S_2$ appears on the right side of Eq. (126), it does not contribute to the relations derived below because its contributions are of higher order in ϵ (and exhibit a second-harmonic dependence on the eikonal phase).

4.5.1. Linear polarization and magnetization. The linear displacement $\boldsymbol{\rho}_\epsilon = \epsilon\,\boldsymbol{\xi}$ is eikonal-phase dependent and its amplitude is $\widetilde{\boldsymbol{\rho}}_\epsilon = \epsilon\,\widetilde{\boldsymbol{\xi}}$. Hence, the push-forward formula (59) yields the eikonal-amplitude of the linear polarization

$$\widetilde{\mathbf{P}}_1 = \sum e\,\overline{n}\,\overline{\left[\widetilde{\boldsymbol{\xi}}\right]} = -\sum \overline{n}\,\frac{\partial\overline{[\Psi_{20}]}}{\partial\widetilde{\mathbf{E}}_1^*}, \tag{127}$$

where $\overline{[\cdots]}$ denotes a momentum-space average over the oscillation-center distribution $\overline{F}$.

The push-forward formula (63), on the other hand, yields the linear expression for the eikonal-dependent current

$$\widetilde{\mathbf{J}}_1 = -i\omega\,\widetilde{\mathbf{P}}_1 + i\,\mathbf{k}\times\left(\sum e\,\overline{n}\,\overline{\left[\widetilde{\boldsymbol{\xi}}\times\overline{\mathbf{v}}\right]}\right) = -i\omega\,\widetilde{\mathbf{P}}_1 + i\,\mathbf{k}c\times\widetilde{\mathbf{M}}_1$$

so that the linear magnetization is

$$\widetilde{\mathbf{M}}_1 = \sum e\,\overline{n}\,\overline{\left[\widetilde{\boldsymbol{\xi}}\times\frac{\overline{\mathbf{v}}}{c}\right]} = -\sum \overline{n}\,\frac{\partial\overline{[\Psi_{20}]}}{\partial\widetilde{\mathbf{B}}_1^*}, \tag{128}$$

where we used the lowest-order expression $\dot{\overline{\mathbf{x}}}_\epsilon = \overline{\mathbf{v}}$ for the reduced particle velocity. Note that the first-order magnetization (128) only has a moving electric-dipole contribution (i.e., $\widetilde{\boldsymbol{\mu}}_1 \equiv 0$) since the intrinsic magnetization (65) is quadratic in $\boldsymbol{\rho}_\epsilon$.

4.5.2. Ponderomotive polarization and magnetization. Ponderomotive polarization and magnetization effects enter at second order through eikonal-phase averaging. We begin with the eikonal-averaged reduced displacement

$$\langle\boldsymbol{\rho}_\epsilon\rangle = \epsilon^2\,\mathrm{Re}\left[\,i\,\mathbf{k}\cdot\left(\widetilde{\boldsymbol{\xi}}^*\,\widetilde{\boldsymbol{\xi}}\right) - i\,m\omega'\widetilde{\boldsymbol{\xi}}\cdot\frac{\partial\widetilde{\boldsymbol{\xi}}^*}{\partial\mathbf{p}}\,\right] = \epsilon^2\,\mathbf{k}\times\left(i\,\widetilde{\boldsymbol{\xi}}\times\widetilde{\boldsymbol{\xi}}^*\right) \equiv \epsilon^2\left(e^{-1}\,\overline{\boldsymbol{\pi}}_2\right),$$

from which we recover the ponderomotive electric-dipole moment (96). Hence, the push-forward relation (59) yields the second-order ponderomotive polarization

$$\overline{\mathbf{P}}_2 = \epsilon^{-2}\sum e\,\overline{n}\,\overline{[\langle\boldsymbol{\rho}_\epsilon\rangle]} = \sum \overline{n}\,\overline{[\overline{\boldsymbol{\pi}}_2]} \equiv -\sum \overline{n}\,\frac{\partial\overline{[\Psi_{21}]}}{\partial\mathbf{E}_0}. \tag{129}$$

We note that the quadrupolar contribution to the ponderomotive polarization (129) is of the form $\epsilon^2\,\nabla\cdot\overline{\mathsf{Q}}_2$, where the ponderomotive electric quadrupole moment (tensor) is defined as $\overline{\mathsf{Q}}_2 \equiv \sum e\overline{n}\;\mathrm{Re}(\overline{[\widetilde{\boldsymbol{\xi}}\,\widetilde{\boldsymbol{\xi}}^*]})$. This contribution is one order higher in the eikonal analysis than the ponderomotive electric-dipole polarization (i.e., $|\nabla\cdot\overline{\mathsf{Q}}_2| \sim \epsilon_0\,|\overline{\mathbf{P}}_2|$) and the ponderomotive quadrupolar polarization is therefore omitted.

Next, we derive an expression for the second-order ponderomotive magnetization. First, the moving-dipole contribution is $\langle(e\,\boldsymbol{\rho}_\epsilon\times\dot{\overline{\mathbf{x}}}_\epsilon)\rangle = e\,\langle\boldsymbol{\rho}_\epsilon\rangle\times\dot{\overline{\mathbf{x}}}_\epsilon = \epsilon^2\;\overline{\boldsymbol{\pi}}_2\times\overline{\mathbf{v}}$ to lowest order in ϵ. Second, the intrinsic ponderomotive magnetization is

$$\left\langle\left(\frac{e}{2}\,\boldsymbol{\rho}_\epsilon\times\frac{d_\epsilon\boldsymbol{\rho}_\epsilon}{dt}\right)\right\rangle = e\,\epsilon^2\,\omega'\left(i\,\widetilde{\boldsymbol{\xi}}\times\widetilde{\boldsymbol{\xi}}^*\right) \equiv \epsilon^2\,c\,\overline{\boldsymbol{\mu}}_2,$$

where we used $(\boldsymbol{\rho}_\epsilon, d_\epsilon\boldsymbol{\rho}_\epsilon/dt) \rightarrow (\epsilon\,\widetilde{\boldsymbol{\xi}},\; -i\epsilon\,\omega'\widetilde{\boldsymbol{\xi}})$. The push-forward relation (64) yields the second-order ponderomotive magnetization

$$\begin{aligned}\overline{\mathbf{M}}_2 &= \epsilon^{-2}\sum\frac{e}{c}\,\overline{n}\,\overline{\left[\left\langle\boldsymbol{\rho}_\epsilon\times\left(\frac{1}{2}\,\frac{d_\epsilon\boldsymbol{\rho}_\epsilon}{dt} + \frac{d_\epsilon\overline{\mathbf{x}}}{dt}\right)\right\rangle\right]} = \sum\overline{n}\,\overline{\left[\overline{\boldsymbol{\mu}}_2 + \overline{\boldsymbol{\pi}}_2\times\frac{\overline{\mathbf{v}}}{c}\right]} \\ &\equiv -\sum\overline{n}\,\frac{\partial\overline{[\Psi_{21}]}}{\partial\mathbf{B}_0}.\end{aligned} \tag{130}$$

We note that the ponderomotive magnetization contribution to the ponderomotive polarization is obtained from the relativistic correction (66) as

$$\begin{aligned} e \left\langle \boldsymbol{\rho}_\epsilon \, \mathsf{T}_\epsilon^{-1}(\gamma - 1) \right\rangle &= \frac{e}{c^2} \left\langle \boldsymbol{\rho}_\epsilon \left(\overline{\mathbf{v}} \cdot \frac{d_\epsilon \boldsymbol{\rho}_\epsilon}{dt} \right) \right\rangle + \cdots \\ &= \epsilon^2 \left[\frac{e}{c^2} \, \overline{\mathbf{v}} \times \left(i\,\omega' \, \widetilde{\boldsymbol{\xi}} \times \widetilde{\boldsymbol{\xi}}^* \right) \right] \equiv \frac{\overline{\mathbf{v}}}{c} \times \left(\epsilon^2 \, \overline{\boldsymbol{\mu}}_2 \right), \end{aligned} \tag{131}$$

where $d_\epsilon \overline{\mathbf{x}}/dt \equiv \overline{\mathbf{v}}$ (to lowest order in ϵ) and we have omitted the relativistic correction $(|\overline{\mathbf{v}}|^2/2c^2)\langle \boldsymbol{\rho}_\epsilon \rangle$ to the ponderomotive polarization (129).

Equations (129) and (130) represent the main results of this Section. They showcase the complementarity of the push-forward and variational methods for the case of the oscillation-center dynamics of charged particles in a weakly-magnetized background plasma.

5. Summary

From a historical perspective, the derivation of oscillation-center Hamiltonian dynamics provided an ideal battleground for the application of Lie-transform perturbation methods in plasma physics, in which the Berkeley School of Plasma Physics, led by Allan N. Kaufman and his collaborators, played a fundamental role. In the spirit of the Berkeley School, the present paper demonstrated the interconnectedness of the Lie-transform perturbation method with the variational approach.

The variational formulation (118) of the *oscillation-center* Vlasov-Maxwell equations provides a simple description of self-consistent ponderomotive polarization and magnetization effects in weakly-magnetized plasmas, derived by variational derivatives with respect to the background electromagnetic fields $(\mathbf{E}_0, \mathbf{B}_0)$. Although the treatment presented here does not include relativistic effects, the variational formalism is naturally suited for relativistic generalization.

Ponderomotive effects for unmagnetized plasmas [11] can be easily recovered by setting $\mathbf{A}_0 \equiv 0$ in the final expressions. Ponderomotive effects in strongly-magnetized plasma, on the other hand, require a two-step transformation: first, a transformation from particle to guiding-center phase space is introduced [4] to be followed by a transformation to oscillation-center phase space [10, 11]. Lastly, low-frequency ponderomotive polarization and magnetization effects in magnetized plasmas are described in terms of the nonlinear gyrokinetic Vlasov-Maxwell equations (e.g., see Ref. [8]) or the nonlinear electromagnetic drift-fluid equations [20].

Acknowledgments

Portions of this paper were presented as an invited talk in celebration of Allan Kaufman's 80th birthday during the KaufmanFest 2007 Symposium in Berkeley, October 6-7, 2007. The Symposium also marked the 50th anniversary of Allan's first paper in plasma physics (with S. Chandrasekhar and K. M. Watson [35]).

Much of the work presented in this paper was done by the Author with support from the U. S. Department of Energy through the Offices of Fusion Energy Sciences and Basic Energy Sciences as well as the National Science Foundation.

Last but not least, Allan's constant support and encouragement over the past 20 years are most gratefully acknowledged. I have learned greatly from working with him and have benefited immensely from his keen physical insights. In fact, the present manuscript benefited greatly from Allan's critical reading and insightful suggestions. I am proud to declare myself a member of the Berkeley School of Plasma Physics.

References

[1] A. N. Kaufman, *The Lie transform: A new approach to classical perturbation theory*, in AIP Conference Proceedings **46**, 286 (1978).

[2] J. R. Cary, Phys. Rep. **79**, 129 (1981).
[3] R. G. Littlejohn, J. Math. Phys. **23**, 742 (1982).
[4] R. G. Littlejohn, J. Plasma Phys. **29**, 111 (1983).
[5] A. N. Kaufman, Phys. Fluids **29**, 1736 (1986).
[6] For a recent historical survey, see J. R. Cary and A. J. Brizard, *Hamiltonian theory of guiding-center motion*, to appear in Reviews of Modern Physics (2009).
[7] A. N. Kaufman and B. M. Boghosian, Contemp. Math. **28**, 169 (1984).
[8] A. J. Brizard and T. S. Hahm, Rev. Mod. Phys. **79**, 421 (2007).
[9] S. Johnston and A. N. Kaufman, Phys. Rev. Lett. **40**, 1266 (1978).
[10] C. Grebogi, A. N. Kaufman, and R. G. Littlejohn, Phys. Rev. Lett. **43**, 1668 (1979).
[11] J. R. Cary and A. N. Kaufman, Phys. Fluids **24**, 1238 (1981).
[12] T. Hatori and H. Washimi, Phys. Rev. Lett. **46**, 240 (1981).
[13] A. N. Kaufman and D. D. Holm, Phys. Lett. A **105**, 277 (1984).
[14] P. L. Similon and A. N. Kaufman, Phys. Rev. Lett. **53**, 1061 (1984).
[15] P. L. Similon, Phys. Lett. A **112**, 33 (1985).
[16] P. L. Similon, A. N. Kaufman, and D. D. Holm, Phys. Fluids. **29**, 1908 (1986).
[17] B. M. Boghosian, *Covariant Lagrangian methods of relativistic plasma theory*, Ph. D. thesis, University of California, Davis (1987); see also arXiv:physics/0307148.
[18] For some recent papers on applications of the ponderomotive force to trap and cool classical particles, see: I. Y. Dodin and N. J. Fisch, Phys. Rev. E **77**, 036402 (2008); I. Y. Dodin and N. J. Fisch, Phys. Plasmas **14**, 055901 (2007); I. Y. Dodin and N. J. Fisch, Phys. Lett. A **349**, 356 (2006); and I. Y. Dodin and N. J. Fisch, Phys. Rev. Lett. **95**, 115001 (2005).
[19] A. J. Brizard, Phys. Rev. Lett. **84**, 5768 (2000).
[20] A. J. Brizard, R. E. Denton, B. Rogers, and W. Lotko, Phys. Plasmas **15**, 082302 (2008).
[21] For a few representative papers by R. L. Dewar, see: R. L. Dewar, Phys. Fluids **13**, 2710 (1970); and R. L. Dewar, Phys. Fluids **16**, 1102 (1973).
[22] For a few representative papers by P. J. Morrison, see: P. J. Morrison and J. M. Greene, Phys. Rev. Lett. **45**, 790 (1980); P. J. Morrison, Rev. Mod. Phys. **70**, 467 (1998); and P. J. Morrison, Phys. Plasmas **12**, 058102 (2005).
[23] For a few representative papers by J. A. Krommes, see: D. H. E. Dubin, J. A. Krommes, C. Oberman, and W. W. Lee, Phys. Fluids **26**, 3524 (1983); and J. A. Krommes, Phys. Rep. **360**, 1 (2002).
[24] A. J. Brizard, Communications in Nonlinear Science and Numerical Simulation, **13**, 24 (2008).
[25] See, for example, A. J. Brizard, J. Plasma Phys. **71**, 225 (2005).
[26] H. Ye and A. N. Kaufman, Phys. Fluids B **4**, 1735 (1992).
[27] J. D. Jackson, *Classical Electrodynamics, 2nd ed.* (Wiley, New York, 1975), chap. 6.
[28] R. N. C. Pfeifer, T. A. Nieminen, N. R. Heckenberg, and H. Rubinsztein-Dunlop, Rev. Mod. Phys. **79**, 1197 (2007).
[29] Y. N. Obukhov, Ann. Phys. (Berlin) **17**, 830 (2008).
[30] P. P. Sosenko, P. Bertrand, and V. Decyk, Phys. Scr. **64**, 264 (2001).
[31] See for example G. P. Fisher, Am. J. Phys. **39**, 1528 (1971).
[32] A. J. Brizard, Phys. Plasmas **7**, 4816 (2000).
[33] J. R. Cary and A. N. Kaufman, Phys. Rev. Lett. **39**, 402 (1977).
[34] A. N. Kaufman, Phys. Rev. A **36**, 982 (1987).
[35] S. Chandrasekhar, A. N. Kaufman, and K. Watson, Ann. Phys. **2**, 435 (1957).

Journal of Physics: Conference Series **169** (2009) 012004 doi:10.1088/1742-6596/169/1/012004

MHD Memes

R L Dewar,[1,2] **R Mills**[1] **and M J Hole**[1]

[1]Department of Theoretical Physics and Plasma Research Laboratory, Research School of Physics and Engineering, The Australian National University, Canberra ACT 0200, Australia
[2]Department of Advanced Energy, Graduate School of Frontier Sciences, University of Tokyo, Kashiwa-City Kashiwanoha 5-1-5 Chiba 277-8561, Japan

E-mail: robert.dewar@anu.edu.au

Abstract. The celebration of Allan Kaufman's 80th birthday was an occasion to reflect on a career that has stimulated the mutual exchange of ideas (or memes in the terminology of Richard Dawkins) between many researchers. This paper will revisit a meme Allan encountered in his early career in magnetohydrodynamics, the continuation of a magnetohydrodynamic mode through a singularity, and will also mention other problems where Allan's work has had a powerful cross-fertilizing effect in plasma physics and other areas of physics and mathematics. To resolve the continuation problem we regularize the Newcomb equation, solve it in terms of Legendre functions of imaginary argument, and define the small weak solutions of the Newcomb equation as generalized functions in the manner of Lighthill, i.e. via a limiting sequence of analytic functions that connect smoothly across the singularity.

1. Introduction

Richard Dawkins [1, 2], in discussing the evolution of ideas, introduced the concept of *memes* in the following terms "... new replicators are not DNA ... patterns of information that can thrive only in brains ... or books, computers and so on ... called memes to distinguish them from genes ... passed from brain to brain ... As they propagate they can change mutate ... memic evolution ..."

Applying and extending this biological analogy to the evolution of science, we can regard scientific papers as the "organisms" in which multiple memes are expressed, and say that cross-fertilization leads to hybrid vigour.

One of Allan Kaufman's greatest contributions to plasma physics has been his role in adapting and developing powerful theoretical and mathematical methods, applying them to plasma physics problems, and propagating these ideas to following generations through his ability as a great teacher.

My (RLD) research interests have intersected, and continue to intersect, with Allan's varied research interests. In this paper I have chosen to illustrate this with a few examples from magnetohydrodynamics (MHD). This may seem a little unexpected, because Allan is not usually thought of as an MHD theorist, but in section 2 I revisit a problem he was familiar with in his early career, namely the continuation (or lack thereof) of solutions of the Newcomb equation for linearized ideal marginal MHD modes. The disconnection between solutions on either side of a mode rational surface is a tricky point to explain to newcomers in the field and I present a new approach which I hope makes this phenomenon clearer.

In section 3 I mention some other MHD problems I have encountered which have intersections with Allan's research.

2. The connection problem at a singular point of the Newcomb equation

It is probably not widely known that some of Allan's earliest work was on magnetohydrodynamic (MHD) stability theory [3]. Newcomb's famous paper on ideal-MHD stability theory in cylindrical geometry [4] acknowledges Allan's critical reading of the manuscript while his Ref. 24 is the note "M. N. ROSENBLUTH (private communication through A. N. Kaufman)," showing that Allan's cross-pollinating role in plasma physics started very early on.

It happens that these papers are highly relevant to our current research [5, 6, 7] on equilibrium and stability in a multiple-region relaxed-MHD plasma model. This is motivated by the desire to construct a well-posed three-dimensional MHD equilibrium theory, based on Taylor relaxation in regions separated by arbitrarily thin ideal-MHD toroidal surfaces that act as barriers to field-line chaos. Not only does this involve the Hamiltonian nonlinear dynamics of the field-line flow, but the Hamiltonian of the field lines must itself be determined self-consistently using MHD theory.

To understand the nature of these ideal-MHD barriers better we have recently [8, 9] returned to the cylindrical limit and looked at the problem of whether the assumed barrier between two relaxed regions of different pressure can be constructed as the zero-width limit of a finite-width ideal-MHD region, stable with a suitably chosen physical pressure profile. [The criterion for physicality of the pressure profile is that the pressure $P(r)$ must be non-negative, where r is the distance from the z-axis.] This problem was first studied by Newcomb and Kaufman [3, 4].

The "Newcomb equation" [4] is satisfied at marginal stability (i.e. when the growth rate γ is zero) by $\xi(r) \equiv \boldsymbol{\xi}\cdot\mathbf{e}_r$, where $\boldsymbol{\xi} \propto \exp i(m\theta + kz)$ is the plasma displacement away from equilibrium, θ being the angle about the z-axis and $\mathbf{e}_r$ the unit vector in the radial direction:

$$\frac{d}{dr}\left(f\,\frac{d\xi}{dr}\right) - g\,\xi = 0, \tag{1}$$

where

$$f(r) \equiv \frac{r[mB_\theta(r) + krB_z(r)]^2}{k^2r^2 + m^2}, \tag{2}$$

$$g(r) \equiv \frac{2k^2r^2}{k^2r^2+m^2}\frac{dP}{dr} + \frac{[mB_\theta(r) + krB_z(r)]^2(k^2r^2+m^2-1)}{r(k^2r^2+m^2)} + \frac{2k^2r[k^2r^2B_z(r)^2 - m^2B_\theta(r)^2]}{(k^2r^2+m^2)^2}. \tag{3}$$

For given integers m and $n \equiv -Rk \neq 0$, a *mode rational surface* is one where $q(r_s) = m/n$, with $q(r) \equiv rB_z/RB_\theta$, B_θ and B_z being respectively the z and θ components of the magnetic field, and R being the periodicity length of the cylinder. At such a rational surface r_s, (1) has a *singular point* because the coefficient of the highest derivative, $f(r)$, vanishes there. Defining $x \equiv r - r_s$, the Taylor expansions of f and g about r_s are of the form $f(r) = f''(r_s)x^2/2 + O(x^3)$ and $g(r) = g(r_s) + O(x)$, with $f''(r_s) > 0$ provided $q'(r_s) \neq 0$ and $g(r_s) \neq 0$ provided $P'(r_s) \neq 0$. The general solution of (1) on intervals not including r_s is found as a linear superposition of two functions whose Frobenius expansions (see e.g. [10, 11]) begin with the fractional powers $|x|^{-1/2\pm\mu}$, where

$$\mu \equiv \frac{\sqrt{1 + 8g(r_s)/f''(r_s)}}{2}. \tag{4}$$

We assume μ is real, otherwise the system is interchange-unstable by the Suydam criterion. The solutions with leading terms $|x|^{-1/2-\mu}$ and $|x|^{-1/2+\mu}$ were called by Newcomb the *large* and *small* solutions, respectively, the large solution being dominant as $x \to 0$.

In this paper we discuss a point raised by Newcomb [4, p. 241]: "In general it is not possible to continue an Euler–Lagrange solution past a singular point." To this he adds the footnote: "This is true only because ξ is a real variable. If it were complex we could go around the singular point by analytic continuation, but the resulting solutions would generally be multivalued." The issue is interesting in the context of the present paper because this meme is perhaps expressed in a mutated form in Allan's later papers on mode conversion (e.g. [12]), which also involves continuation through singularities.

The natural framework in which to understand the singular solutions of the Newcomb equation is generalized function theory because the solutions are not defined pointwise (being undefined at $r = r_s$). In our previous work [13, 14, 15], using the singular solutions of the Newcomb equation or the two-dimensional generalization of it due to Bineau [16], we have followed the approach and notation of of Gel'fand and Shilov [17] based on inner products with all members of a space of smooth test functions.

However this approach is rather abstract and difficult to visualize, so in the current paper we present an alternative approach, where the required generalized functions are defined as limits of sequences, parametrized by δ, of smooth functions that tend to weak solutions of the Newcomb equation as $\delta \to +0$. This is in the spirit of Lighthill's [18] approach to generalized function theory.

To do this we study a model Newcomb equation, in which $f(x) = x^2$ and $g(x) = \mu^2 - 1/4$. The weak solution space is spanned by the *four* Gel'fand-Shilov generalized functions

$$\begin{aligned}
\Xi_-^{\rm Fr}(x) &\equiv |x|^{-1/2-\mu}\,{\rm sgn}\,x \\
\Xi_+^{\rm Fr}(x) &\equiv |x|^{-1/2-\mu} \\
\xi_-^{\rm Fr}(x) &\equiv |x|^{-1/2+\mu}\,{\rm sgn}\,x \\
\xi_+^{\rm Fr}(x) &\equiv |x|^{-1/2+\mu},
\end{aligned} \tag{5}$$

where ${\rm sgn}\,x$ denotes the sign of x. On the left-hand sides we have used the notation of Dewar and Persson [11] in which the large and small Frobenius solutions are indicated by using upper and lower case greek, respectively; odd and even parity solutions are indicated by subscript $-$ and $+$, respectively.

Alternatively, we may use the "single-sided" generalized functions $|x|_\pm^\lambda \equiv \frac{1}{2}(|x|^\lambda \pm |x|^\lambda\,{\rm sgn}\,x)$ whose support is the positive ($x > 0$) or negative ($x < 0$) half line. When restricted to their domains of support these single-sided solutions correspond to the classical pointwise solutions Newcomb meant when he said solutions of the Newcomb equation could not be continued through the singular point. As the domains of support are to the left (L) and right (R) of the origin, we denote the corresponding generalized function solutions of the Newcomb equation by subscripts L and R

$$\begin{aligned}
\Xi_{\rm L}^{\rm Fr}(x) &\equiv |x|_-^{-1/2-\mu} \\
\Xi_{\rm R}^{\rm Fr}(x) &\equiv |x|_+^{-1/2-\mu} \\
\xi_{\rm L}^{\rm Fr}(x) &\equiv |x|_-^{-1/2+\mu} \\
\xi_{\rm R}^{\rm Fr}(x) &\equiv |x|_+^{-1/2+\mu}.
\end{aligned} \tag{6}$$

In ideal MHD the large solutions are rejected as they are not square integrable (hence give infinite kinetic energy), thus reducing the dimensionality of the solution space to 2 as expected for second-order differential equations. (However, the large solutions are essential in resistive stability theory for matching asymptotically to the resistive internal layer in the neighbourhood of the singular point.)

To apply the Lighthill approach we use a regularization appropriate to ideal-MHD stability studies, in which the only regularizing effect is that of inertia from the mass density ρ. This comes into play when the growth rate γ is nonzero, adding to f a positive term proportional to $\rho\gamma^2$ (see e.g. [19]) and thus removing the singularity but approximating the Newcomb equation arbitrarily closely as $\gamma \to 0$. Thus we study the *regularized model Newcomb equation*

$$\frac{d}{dx}(x^2+\delta^2)\frac{d}{dx}\xi - \left(\mu^2 - \frac{1}{4}\right)\xi = 0, \tag{7}$$

whose general solution is a linear combination of the respectively odd and even functions $\operatorname{Im} P_{\mu-1/2}(ix/\delta)$ and $\operatorname{Re} P_{\mu-1/2}(ix/\delta)$, where $P_\nu(z) \equiv P_\nu^0(z)$ denotes a Legendre function of the first kind [20, chapter 8].

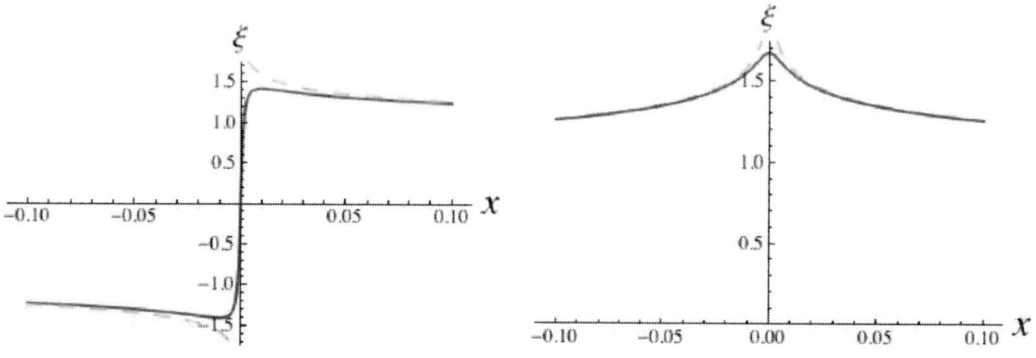

Figure 1. Left panel: regularized odd solution $\xi_-^{\rm Fr}(x|0.001)$ (solid line) and unregularized odd solution $\xi_-^{\rm Fr}(x)$ (dashed line). Right panel: regularized even solution $\xi_+^{\rm Fr}(x|0.005)$ (solid line) and unregularized even solution $\xi_+^{\rm Fr}(x)$ (dashed line). In both cases $\mu = 0.4$.

By expanding $P_{\mu-1/2}(ix/\delta)$ in powers of δ and choosing the superposition coefficients appropriately we can now define Lighthill approximating functions for the odd and even small solutions

$$\xi_-^{\rm Fr}(x|\delta) \equiv \frac{(2\delta)^{\mu-\frac{1}{2}}\Gamma\left(\mu+\frac{1}{2}\right)^2}{\sin\left(\frac{\pi}{2}\left(\mu-\frac{1}{2}\right)\right)\Gamma(2\mu)} \operatorname{Im} P_{\mu-\frac{1}{2}}\left(\frac{ix}{\delta}\right), \tag{8}$$

$$\xi_+^{\rm Fr}(x|\delta) \equiv \frac{(2\delta)^{\mu-\frac{1}{2}}\Gamma\left(\mu+\frac{1}{2}\right)^2}{\cos\left(\frac{\pi}{2}\left(\mu-\frac{1}{2}\right)\right)\Gamma(2\mu)} \operatorname{Re} P_{\mu-\frac{1}{2}}\left(\frac{ix}{\delta}\right). \tag{9}$$

Figure 1 shows examples of the unregularized solutions defined in (5) and the regularized solutions defined in (8) and (9).

The leading terms of the asymptotic expansions of these two functions in powers of $\delta^{2\mu}$ are

$$\begin{aligned} \xi_-^{\rm Fr}(x|\delta) &= |x|^{\mu-\frac{1}{2}}\operatorname{sgn} x + \delta^{2\mu}c_-(\mu)\,|x|^{-\mu-\frac{1}{2}}\operatorname{sgn} x + O(\delta^{4\mu}), \\ \xi_+^{\rm Fr}(x|\delta) &= |x|^{\mu-\frac{1}{2}} + \delta^{2\mu}c_+(\mu)\,|x|^{-\mu-\frac{1}{2}} + O(\delta^{4\mu}), \end{aligned} \tag{10}$$

where the factors $c_\pm(\mu)$ are defined by

$$\begin{aligned} c_-(\mu) &\equiv -2^{2\mu}\frac{\Gamma\left(\mu+\frac{1}{2}\right)^2\Gamma(-2\mu)}{\Gamma\left(\frac{1}{2}-\mu\right)^2\Gamma(2\mu)}\frac{\sin\left(\frac{\pi}{2}\left(\mu+\frac{1}{2}\right)\right)}{\sin\left(\frac{\pi}{2}\left(\mu-\frac{1}{2}\right)\right)} \\ c_-(\mu) &\equiv 2^{2\mu}\frac{\Gamma\left(\mu+\frac{1}{2}\right)^2\Gamma(-2\mu)}{\Gamma\left(\frac{1}{2}-\mu\right)^2\Gamma(2\mu)}\frac{\cos\left(\frac{\pi}{2}\left(\mu+\frac{1}{2}\right)\right)}{\cos\left(\frac{\pi}{2}\left(\mu-\frac{1}{2}\right)\right)}. \end{aligned} \tag{11}$$

The $O(1)$ terms are the odd and even generalized function small solutions as required. The $O(\delta^{2\mu})$ terms involve the large solutions of the same parity, with coefficients that vanish as $\delta \to +0$ because our regularization is based on ideal MHD.

The special case $\mu \to 1/2$ is important because it corresponds to the zero-β limit, or flattening pressure at a rational surface at arbitrary β. In this limit we find

$$\begin{aligned} \xi_-^{\rm Fr}(x|\delta) &= \operatorname{sgn} x - \frac{2\delta}{\pi} x^{-1} + O(\delta^2), \\ \xi_+^{\rm Fr}(x|\delta) &= 1 + O(\delta^2). \end{aligned} \tag{12}$$

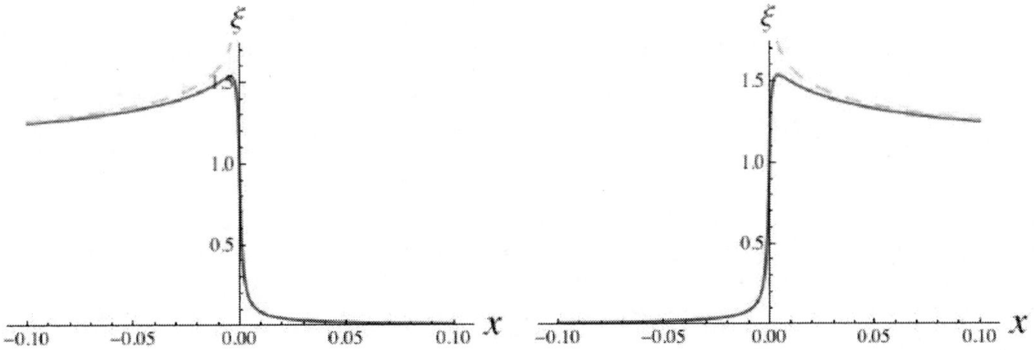

Figure 2. Left panel: regularized left solution $\xi_{\rm L}^{\rm Fr}(x|0.001)$ (solid line) and unregularized left solution $\xi_{\rm L}^{\rm Fr}(x)$ (dashed line). Right panel: regularized right solution $\xi_{\rm R}^{\rm Fr}(x|0.001)$ (solid line) and unregularized right solution $\xi_{\rm R}^{\rm Fr}(x)$ (dashed line). In both cases $\mu = 0.4$.

We are now ready to regularize Newcomb's "disconnected" solutions by combining the regularized odd and even parity solutions in an analogous way to that used for unregularized solutions in (6),

$$\xi_{\rm L}^{\rm Fr}(x|\delta) \equiv \frac{1}{2}\left[\xi_+^{\rm Fr}(x|\delta) - \xi_-^{\rm Fr}(x|\delta)\right], \tag{13}$$

$$\xi_{\rm R}^{\rm Fr}(x|\delta) \equiv \frac{1}{2}\left[\xi_+^{\rm Fr}(x|\delta) + \xi_-^{\rm Fr}(x|\delta)\right], \tag{14}$$

so that the $O(\delta^0)$ terms cancel to the right or left of the origin for $\xi_{\rm L}^{\rm Fr}(x|\delta)$ and $\xi_{\rm R}^{\rm Fr}(x|\delta)$, respectively.

Figure 2 shows examples of the unregularized solutions defined in (6) and the regularized solutions defined in (13) and (14). It is seen that, for any finite δ, the solutions do connect across the singular point at the origin. By using (10) in (13) and (14) we see that the L and R solutions decay rapidly (like the large solution) as $x \to \pm\infty$ with an amplitude that tends to zero as $\delta \to +0$, so that their support becomes the negative or positive half line in the limit.

3. Other MHD intersections

The most obvious point of intersection between Allan Kaufman's and my (RLD) research interests is in the oscillation centre theory and the theory of wave action, which goes back to an MHD paper [21]. This has developed in various ways that other contributors to this Proceedings will no doubt touch on in much more detail. However, sticking to the MHD theme, it could be said that the theory of flux-minimizing surfaces [22, 23, 24, 25, 26] is a mutant form of oscillation centre theory, and this has close links with our current research on three-dimensional MHD equilibrium theory [5, 6].

Another intersecton concerns the theory of quantum chaos, a field in which Allan made an important contribution [27], which is better known outside the field of plasma physics than within it. Indeed there has been very little in the plasma physics literature about quantum chaos, perhaps because of the name. However, this field is really about WKB theory for arbitrary waves in the case of non-integrable ray dynamics, and in a number of papers over the past few years we have applied methods from this field to the study of the ideal-MHD spectrum in a stellarator [28, 19, 29].

4. Conclusion

In this brief paper we have revisited a basic problem in the the theory of ideal MHD stability theory that has connections with Allan Kaufman's early research and our current research. In his long career Allan has been a role model and inspiration to many, and particularly to the first author of this paper.

Acknowledgements

The first author (RLD) thanks the organizers of KaufmanFest for giving him the opportunity to help celebrate Allan Kaufman's life and contributions. The algebraic manipulations and plots in this paper were done using *Mathematica* [30]. Some of the work mentioned was supported by the Australian Research Council (ARC) Discovery Project DP0452728 and attendance at KaufmanFest was made possible partly by support from the ARC Complex Open Systems Research Network (COSNet).

References

[1] Dawkins R 1976 *The Selfish Gene* (Oxford, UK: Oxford University Press) URL `http://en.wikipedia.org/wiki/The_Selfish_Gene`
[2] Dawkins R 1986 *The Blind Watchmaker* (Harlow, UK: Longman Scientific and Technical)
[3] Newcomb W A and Kaufman A N 1961 *Physics of Fluids* **4** 314 URL `http://link.aip.org/link/?PFL/4/314/1`
[4] Newcomb W A 1960 *Annals of Physics* **10** 232 URL `http://www.sciencedirect.com/science/article/B6WB1-4DDR40K-T1/2/72496b9dddbc90fe5a2f659bfbef1ea6`
[5] Hudson S R, Hole M J and Dewar R L 2007 *Phys. Plasmas* **14** 052505
[6] Hole M J, Hudson S R and Dewar R L 2007 *Nucl. Fusion* **47** 746 URL `http://stacks.iop.org/0029-5515/47/746`
[7] Dewar R L, Hole M J, McGann M, Mills R and Hudson S R 2008, accepted for publication in *Entropy* URL `http://www.mdpi.com/journal/entropy/`
[8] Mills R 2007 *3D Plasma Equilibria* Honours thesis The Australian National University Canberra ACT 0200, Australia
[9] Mills R, Hole M J and Dewar R L 2009 to be submitted to J. Plasma Phys.
[10] Ince E L 1956 *Ordinary Differential Equations* (New York: Dover)
[11] Dewar R L and Persson M 1993 *Phys. Fluids B* **5** 4273
[12] Kaufman A N and Friedland L 1987 *Physics Letters A* **123** 387 URL `http://www.sciencedirect.com/science/article/B6TVM-46MM1XC-3R/2/dfe74ac0b228ec199e63cbd9c7346c10`
[13] Manickam J, Grimm R C and Dewar R L 1981 *Comp. Phys. Comm.* **24** 355
[14] Grimm R C, Dewar R L and Manickam J 1983 *J. Comput. Phys.* **49** 94
[15] Dewar R L and Pletzer A 1990 *J. Plasma Phys.* **43**
[16] Bineau M 1966 *Nucl. Fusion* **2** 130
[17] Gel'fand I M and Shilov G E 1964 *Generalized Functions* (New York: Academic Press)
[18] Lighthill M J 1958 *Introduction to Fourier Analysis and Generalised Functions* (Cambridge University Press)
[19] Dewar R L, Tatsuno T, Yoshida Z, Nührenberg C and McMillan B F 2004 *Phys. Rev. E* **70** 066409
[20] Abramowitz M and Stegun I A (eds) 1972 *Handbook of Mathematical Functions* 10th ed Applied Mathematics Series - 55 (U.S. Government Printing Office, Washington D.C.: National Bureau of Standards) URL `http://www.math.sfu.ca/~cbm/aands/`

[21] Dewar R L 1970 *Phys. Fluids* **11**
[22] Dewar R L and Meiss J D 1992 *Physica D* **57** 476
[23] Dewar R L, Hudson S R and Price P F 1994 *Phys. Lett. A* **194** 49
[24] Hudson S R and Dewar R L 1997 *Phys. Lett. A* **226** 85
[25] Hudson S R and Dewar R L 1998 *Phys. Lett. A* **247** 246
[26] Hudson S R and Dewar R L 1999 *Phys. Plasmas* **6** 1532
[27] McDonald S W and Kaufman A N 1979 *Phys. Rev. Lett.* **42** 1189
[28] Dewar R L, Cuthbert P and Ball R 2001 *Phys. Rev. Letters* **86** 2321
[29] Dewar R L, Kenny B G, Nührenberg C, Tatsuno T and McMillan B F 2007 *J. Korean Phys. Soc.* **50** 112 URL http://www.kps.or.kr/home/kor/journal/library/journal.asp?journaluid={F7A5B5DE-52C4-42EE-B398-E2D737D706EC}&globalmenu=4&localmenu=11&journaltype={C017611D-EF29-4B30-8714-B9265F5AFF31}
[30] Wolfram Research, Inc 2007 *Mathematica, Version 6* (Champaign, Illinois, USA) URL http://support.wolfram.com/mathematica/reference/general/citing.html

Plasma Theory, Wave Kinetics, Nonlinear Dynamics: KaufmanFest 2007 IOP Publishing
Journal of Physics: Conference Series **169** (2009) 012005 doi:10.1088/1742-6596/169/1/012005

Nonlinear theoretical tools for fusion-related microturbulence: Historical evolution, and recent applications to stochastic magnetic fields, zonal-flow dynamics, and intermittency

J A Krommes

Plasma Physics Laboratory, Princeton University,
P.O. Box 451, MS28, Princeton, NJ 08543–0451 USA

Abstract. Fusion physics poses an extremely challenging, practically complex problem that does not yield readily to simple paradigms. Nevertheless, various of the theoretical tools and conceptual advances emphasized at the KaufmanFest 2007 have motivated and/or found application to the development of fusion-related plasma turbulence theory. A brief historical commentary is given on some aspects of that specialty, with emphasis on the role (and limitations) of Hamiltonian/symplectic approaches, variational methods, oscillation-center theory, and nonlinear dynamics. It is shown how to extract a renormalized ponderomotive force from the statistical equations of plasma turbulence, and the possibility of a renormalized K–χ theorem is discussed. An unusual application of quasilinear theory to the problem of plasma equilibria in the presence of stochastic magnetic fields is described. The modern problem of zonal-flow dynamics illustrates a confluence of several techniques, including (i) the application of nonlinear-dynamics methods, especially center-manifold theory, to the problem of the transition to plasma turbulence in the face of self-generated zonal flows; and (ii) the use of Hamiltonian formalism to determine the appropriate (Casimir) invariant to be used in a novel wave-kinetic analysis of systems of interacting zonal flows and drift waves. Recent progress in the theory of intermittent chaotic statistics and the generation of coherent structures from turbulence is mentioned, and an appeal is made for some new tools to cope with these interesting and difficult problems in nonlinear plasma physics. Finally, the important influence of the intellectually stimulating research environment fostered by Prof. Allan Kaufman on the author's thinking and teaching methodology is described.

1. Introduction

This article records and expands upon the material I presented at the symposium on *Plasma Theory, Wave Kinetics, and Nonlinear Dynamics: KaufmanFest 2007.* Its goal is to discuss a very biased selection of analytical tools for the study of possibly strong plasma turbulence (especially of the kind relevant to fusion applications) in light of some of the methods pioneered and championed by Prof. Allan Kaufman and his coworkers. Those were inspirational to me early in my career and should be part of the repertoire of every plasma theorist, especially those just getting started. In particular, I will mention multiple-scale techniques [1], quasilinear theory [2, 3], oscillation-center transformations [4, 5], the K–χ theorem [6], Lie perturbation theory [5], and the Weyl calculus [7]. These are very well suited for applications to weak-turbulence theory. In most cases they have not been successfully applied to strong turbulence;

Journal of Physics: Conference Series **169** (2009) 012005 doi:10.1088/1742-6596/169/1/012005

	— Neutral fluids/general physics —		— Plasmas —
1959	**Kraichnan [17]**—Direct-interaction approximation	**1959**	
1960		**1960**	**Balescu [25], Lenard [26]**—BL collision operator
1	**Wyld [20]**—Diagrammatic formulation of NS turbulence	**1**	
·	**Kraichnan [18]**—Stochastic osc.; random-coupling model	·	
2	**Kubo [21]**—Cumulant expansions	**2**	**Vedenov *et al.* [9]**—QLT
·		·	**Drummond & Pines [10]**—QLT
3	**Lorenz [22]**—Deterministic nonperiodic flow	**3**	
4		**4**	
1965		**1965**	**Kadomtsev [11]**—Monograph on *Plasma Turbulence*
6		**6**	**Dupree [15]**—Unmagnetized RBT
7	**Kraichnan [23]**—Dual cascade	**7**	**Orszag & Kraichnan [19]**—Vlasov DIA
·		·	**Dupree [16]**—RBT for $\boldsymbol{E} \times \boldsymbol{B}$ motions
8		**8**	**Weinstock [27]**—Turbulent diffusion in $\boldsymbol{B}$
1969	**Benney & Newell [13]**—Random wave closures	**1969**	**Sagdeev and Galeev [12]**—Monograph on *Nonlinear Plasma Theory*
·	**Chirikov [24]**—PhD thesis on stochasticity	·	

Figure 1. Chronology of selected research papers on statistical methods [DIA (1959) to 1969]. More detailed versions (with some differences of emphasis) of the timeline figures presented here can be found in Ref. 28.

however, given the complexity of the subject, there really is no substitute for elegant and concise methods that apply in any regime whatsoever. They provide important reference results, and it is illuminating to understand when and why they fail and what generalizations or alternatives have evolved. With regard to the latter, I will mention the Martin–Siggia–Rose (MSR) formalism, statistical moment closures such as the direct-interaction approximation (DIA) and the eddy-damped quasinormal Markovian (EDQNM) theory, and the use of model nonlinear Langevin equations for estimating higher-order statistics. I will also describe some recent calculations that employ bifurcation theory, as well as a novel use of a wave kinetic equation to describe *nonlinear* statistical interactions. The latter analysis makes good use of a Hamiltonian formulation of $\boldsymbol{E} \times \boldsymbol{B}$ advection.

I cannot possibly give an adequate discussion of the evolution of plasma turbulence theory through its approximately half-century of development in the present short article. However, I do hope to provide some perspectives and references that may help the reader organize a vast amount of technical information.[1] Initial foundations were laid in the 1960's (see the timeline in Fig. 1) with the development of quasilinear theory (QLT) [9, 10], weak-turbulence theory (WTT) (see Refs. 11–13 and the references in Ref. 14), and resonance-broadening theory (RBT) [15, 16]. For reference, Kraichnan's pioneering DIA was published in 1959 [17] for strong incompressible Navier–Stokes turbulence and generalized in 1961 [18] to arbitrary quadratically nonlinear dynamical systems. Early plasma theorists were mostly oblivious to the DIA, and a fundamental 1967 article by Orszag and Kraichnan [19] critiquing RBT and describing the DIA for Vlasov turbulence was virtually ignored for about a decade.

The 1970's (see Fig. 2) were a time of great excitement and advances on several fronts, including Davidson's systematic treatise on WTT [14], Dewar's pioneering work on oscillation-center QLT [4], and the work of Smith and Kaufman [37] on stochastic heating. In 1976

[1] Another related article with pedagogical intent is my written account of the four introductory lectures on plasma turbulence that I delivered at the 19th Canberra International Physics Summer School [8].

	— Neutral fluids/general physics —		— Plasmas —
1970	**Kuhn [29]**—*The Structure of Scientific Revolutions*	**1970**	
1		1	
·		·	**Kaufman [2]**—Reformulation of QLT
2		2	**Davidson [14]**—*Methods in Nonlinear Plasma Theory*
·	**Candlestickmaker [30]**—Accurate determination of e	·	**Benford & Thompson [36]**—Probabilistic model of RBT
3	**Martin, Siggia, & Rose [31]**—Renormalized classical field theory	3	**Dewar [4]**—Oscillation center quasilinear theory
4		4	
1975		**1975**	**Smith & Kaufman [37]**—Stochastic acceleration by a single wave
6	**Orszag [32]**—Review article on statistical closure	6	**Krommes & Oberman [38]**—Renormalization and convective cells
·	**Kraichnan [33]**—Theory of eddy viscosity		
7		7	**Cary & Kaufman [6]**—K–χ theorem
·		·	**Johnston & Kaufman [39]**—Mode coupling and K–χ theorem *via* Lie transforms
8	**Treve [34]**—Review: Theory of chaotic motion	8	
·		·	**DuBois & Espedal [40], Krommes [41]**—Plasma DIA
1979	**Rose [35]**—Renormalization of many-particle systems	**1979**	**Littlejohn [42]**—A guiding center Hamiltonian

Figure 2. Chronology of selected research papers (1970 to 1979).

I published some work on the nonlinear theory of convective cells [38], a strong-turbulence calculation that led me to appreciate just how difficult the turbulence problem really was. In an attempt to broaden my horizons, I arranged for several lengthy summer visits to Berkeley, circa 1976–8. There I was introduced to Allan's distinctive style and an immensely pleasurable and stimulating intellectual environment (quite different from the mission-oriented ambience of my home base at Princeton University's Plasma Physics Laboratory). Allan, of course, was into everything. But I was also inspired by meeting (among others) John Cary (working on ponderomotive force *via* Hamiltonian–Lie methods); Shayne Johnston (a former fellow grad student at Princeton, then working with Allan on Hamiltonian methods for ponderomotive effects and nonlinear wave–particle interactions); Allan Lichtenberg (stochasticity); Robert Littlejohn (new tools for single-particle motion); Harry Mynick (beam–plasma interaction); and Gary Smith (stochastic heating). I also participated in the reading group that was dissecting Chirikov's thesis line by line; that provided my first introduction to nonlinear dynamics. Either directly or indirectly, all of those interactions and research topics served to inspire my future research.

Perhaps the most significant direct consequence of my Berkeley interactions was the work on nonlinear Hamiltonian gyrokinetics by Dubin, Krommes, Oberman, and Lee [43], which built on the new techniques being developed by Robert Littlejohn [42]. However, although nonlinear gyrokinetics is absolutely key to the modern development of fusion theory and simulation, the gyrokinetics story has been told frequently [8] and reviewed at length [44, 45], so I shall not discuss it further in this article. Instead, I would like to focus on the significance of some of the other tools about which I learned at Berkeley in the context of the turbulence problem, and also point out some instances in which new tools have been required.

2. Tools for Turbulence

In Secs. 2.1–2.5 I shall give a general discussion of the challenges of a statistical theory of turbulence. The relationship of the abstract formalism to the Berkeley methods may not be immediately apparent. However, the patient reader will be rewarded in Sec. 2.6, where classical calculations relating to ponderomotive force will be reconsidered from the point of view of renormalized statistical turbulence theory. Then in Sec. 2.7 I will describe an application of quasilinear theory to the practical problem of the determination of plasma equilibria in the presence of stochastic magnetic fields. That calculation ties together a variety of themes relating to statistical formalism.

2.1. The statistical turbulence problem

A fundamental signature of turbulence is its unpredictability, i.e., its extreme sensitivity to small changes in initial conditions. Nonlinear dynamics provides rigorous proofs that certain dynamical systems can contain behavior that is as random as a coin toss [46, 47]. Such observations motivate the application of statistical methods to nonlinear PDEs (although this is a definite leap of faith because, as Poincaré explained more than a century ago, regular and stochastic regions are usually intertwined in a very complicated fashion [48, 49]). For definiteness, let a deterministic equation evolve a field $\psi(t)$ from an initial condition ψ_0. Formally, this can be done by the simple artifice of assigning a smooth probability density function (PDF) $P(\psi_0)$ to the initial conditions, thus creating a statistical ensemble. The general goal is then to calculate the fully multivariate probability density functional $P[\psi]$. From such a functional, one can in principle extract correlation functions involving n time points for arbitrary $n \geq 1$. Generalizations of such PDFs enable one to calculate n-point response functions as well.

Although this goal is easily stated, it is virtually impossible to achieve it in practice. The physical difficulty is that the various regions of integrable and chaotic behavior present in the typical nonlinear PDE cannot be distinguished by low-order statistical moments. As a (somewhat contrived) mathematical example, consider one ensemble comprising Gaussianly shaped spikes (examples of coherent structures) uniformly distributed on a large interval of width T, namely $\psi(t) = (2\pi\sigma^2)^{-1/2}\exp[-(t-t_0)^2/2\sigma^2]$ with $P(t_0) = T^{-1}$; and another ensemble whose realizations obey the stochastic oscillator equation [18]

$$\partial_t\widetilde{\psi}(t) + \mathrm{i}\widetilde{\Omega}\widetilde{\psi} = 0, \tag{1}$$

where tilde denotes a random variable and $\widetilde{\Omega}$ is a centered Gaussian random variable with $\langle\widetilde{\Omega}^2\rangle = \beta^2$. The covariances $\langle\delta\psi(t)\delta\psi^*(t')\rangle$ (where $\delta\psi \doteq \widetilde{\psi} - \langle\psi\rangle$ and the angle brackets denote ensemble averaging) can both be made identical to $\exp(-\frac{1}{2}\beta^2\tau^2)$ (where $\tau \doteq t-t'$ and $\doteq$ denotes a definition) with the choices $\beta = (2\pi)^{1/2}T$ and $\sigma = (4\pi)^{-1/2}T^{-1}$ [neglecting corrections of $O(T^{-1})$].

In spite of such difficulties, the traditional analytical approach to the turbulence problem has been the construction of statistical closures based on low-order moments. That subject has been reviewed exhaustively [28, and references therein]. More recently, methods based on the construction of entire PDFs have been explored [50, 51]. Yet another approach is a direct attack using methods of nonlinear dynamics [52]. However, while the latter techniques have considerable promise for systems with small number of degrees of freedom (for some related discussion, see Sec. 3.1), it is unclear how to proceed for realistic systems with huge numbers of excited modes; even if there is a strange attractor embedded in the phase space, its dimension may be very large.

2.2. Statistical observables

The first step in developing a statistical formalism is to define the basic dependent variables. That is nontrivial. One obvious candidate is the mean field $\langle\psi\rangle$, but that frequently vanishes due to symmetry. At the level of two-point functions, in an Eulerian framework there are two natural functions: the correlation function $C(t,t') \doteq \langle\delta\psi(t)\delta\psi(t')\rangle$; and the response function $R(t;t') \doteq \langle\delta\psi(t)/\delta\widehat{\eta}(t')\rangle$, where $\widehat{\eta}$ is a statistically sharp source term added to the right-hand side of the dynamical equation, i.e., $\partial_t\psi + \cdots = \widehat{\eta}(t)$. That $\langle\psi\rangle$ and C are cumulants [21] is

Journal of Physics: Conference Series **169** (2009) 012005 doi:10.1088/1742-6596/169/1/012005

clear; the interpretation of R as a cumulant is more subtle and interesting and follows from the work of MSR [28, 31].

Statistically averaged quantities such as $\langle\psi\rangle$, C, and R are called *statistical observables*. ($\widetilde{\psi}$ itself is called the *primitive amplitude* and is assumed to obey a dynamical evolution equation that is of first order in time and possesses a polynomial, typically quadratic, nonlinearity.) In a cumulant description of turbulence, the goal is to find a system of equations closed in a small number of such low-order observables. That is difficult because it is easy to see that nonlinearity in stochastic quantities couples cumulants through all orders. This is the well-known *statistical closure problem* [53].

2.3. Resonance broadening

For an introductory discussion of some of the difficulties inherent in a formalism appropriate for strong turbulence, I shall focus on the response function R, which is a mean Green's function. In simple cases, it can be interpreted as a conditional probability. [Recall that the PDF of a random variable $\widetilde{\psi}$ can be expressed as $P(\psi) = \langle\delta(\psi - \widetilde{\psi})\rangle$.] Two important basic examples of response functions are (i) the single-particle propagator for free-streaming motion, and (ii) the diffusion Green's function. The former obeys

$$\partial_t g^{(0)}(\boldsymbol{x}, \boldsymbol{v}, t; \boldsymbol{x}', \boldsymbol{v}', t') + \boldsymbol{v} \cdot \boldsymbol{\nabla} g^{(0)} = \delta(t - t')\delta(\boldsymbol{x} - \boldsymbol{x}')\delta(\boldsymbol{v} - \boldsymbol{v}'), \tag{2}$$

the solution of which is

$$g^{(0)}(\boldsymbol{x}, \boldsymbol{v}, t; \boldsymbol{x}', \boldsymbol{v}', t') = H(t - t')\delta(\boldsymbol{x} - \boldsymbol{x}' - \boldsymbol{v}(t - t'))\delta(\boldsymbol{v} - \boldsymbol{v}'). \tag{3}$$

[Here $H(\tau)$ is the Heaviside unit step function that reflects causality.] Nothing is random here; $g^{(0)}$ is its own mean. It is the conditional $\boldsymbol{x}$–$\boldsymbol{v}$ phase-space PDF of finding the particle at $(\boldsymbol{x}, \boldsymbol{v})$ at time t, given that it was at $(\boldsymbol{x}', \boldsymbol{v}')$ at t'. This particular function is singular; its Fourier transform[2] ($\boldsymbol{x} - \boldsymbol{x}' \to \boldsymbol{k}$, $t - t' \to \omega$) is

$$\widehat{g}^{(0)}(\boldsymbol{k}, \boldsymbol{v}, \omega; \boldsymbol{v}') = \frac{\delta(\boldsymbol{v} - \boldsymbol{v}')}{-\mathrm{i}(\omega - \boldsymbol{k} \cdot \boldsymbol{v} + \mathrm{i}\epsilon)}, \tag{4}$$

where the presence of ϵ (a positive infinitesimal) reflects causality. Note for future use that the reduced, $\boldsymbol{v}$-space PDF is $g^{(0)}(\boldsymbol{v}, t; \boldsymbol{v}', t') = \int d\boldsymbol{x}\, g^{(0)}(\boldsymbol{x}, \boldsymbol{v}, t; \boldsymbol{x}', \boldsymbol{v}', t') = H(t - t')\delta(\boldsymbol{v} - \boldsymbol{v}')$ or

$$\widehat{g}^{(0)}(\boldsymbol{v}, \omega; \boldsymbol{v}') = \frac{\delta(\boldsymbol{v} - \boldsymbol{v}')}{-\mathrm{i}(\omega + \mathrm{i}\epsilon)}, \quad \widehat{g}^{(0)}(\boldsymbol{\lambda}, \omega) = \frac{1}{-\mathrm{i}(\omega + \mathrm{i}\epsilon)}, \tag{5}$$

where $\boldsymbol{\lambda}$ is the Fourier wave number conjugate to $\boldsymbol{v} - \boldsymbol{v}'$.[3]

The diffusion Green's function is $g(\boldsymbol{v}, t; \boldsymbol{v}', t') \doteq \langle\widetilde{g}(\boldsymbol{v}, t; \boldsymbol{v}', t')\rangle$, where

$$\partial_t \widetilde{g}(\boldsymbol{v}, t; \boldsymbol{v}', t') + \widetilde{\boldsymbol{a}}(t) \cdot \partial_{\boldsymbol{v}} \widetilde{g} = \delta(t - t')\delta(\boldsymbol{v} - \boldsymbol{v}') \tag{6}$$

and $\widetilde{\boldsymbol{a}}(t) = (q/m)\widetilde{\boldsymbol{E}}(t)$ is Gaussian white noise characterized by $\langle\widetilde{\boldsymbol{a}}\rangle = \boldsymbol{0}$ and $\langle\delta\widetilde{\boldsymbol{a}}(t)\delta\widetilde{\boldsymbol{a}}(t')\rangle = 2D\delta(t - t')$. Although the most conventional way of solving for g is to recognize that the average of Eq. (6) is rigorously the diffusion equation

$$\partial_t g(\boldsymbol{v}, t; \boldsymbol{v}', t') - D\nabla_v^2 g = \delta(t - t')\delta(\boldsymbol{v} - \boldsymbol{v}'), \tag{7}$$

[2] A consistent set of conventions is described in Appendix A of either Ref. 28 or Ref. 45. In this paper I use the circumflex to denote the Fourier transform in at least one of the variables.

[3] Translational invariance in $\boldsymbol{v}$ is not a general property of velocity-dependent Green's functions. For example, $g^{(0)}(\boldsymbol{x}, \boldsymbol{v}, t; \boldsymbol{x}', \boldsymbol{v}', t')$ does not possess that property.

an alternate approach is to note that the solution of Eq. (6) is, via the method of characteristics,

$$\widetilde{g}(\boldsymbol{v},t;\boldsymbol{v}',t') = \delta\left(\boldsymbol{v}-\boldsymbol{v}'-\int_{t'}^{t} d\bar{t}\,\widetilde{\boldsymbol{a}}(\bar{t})\right), \tag{8}$$

the average of which can be calculated using cumulant expansions (which truncate at order 2 for Gaussian statistics). Either way, one finds

$$\widehat{g}(\boldsymbol{v},t;\boldsymbol{v}',t') = H(t-t')\exp\left(-\frac{|\boldsymbol{v}-\boldsymbol{v}'|^2}{4D(t-t')}\right), \tag{9a}$$

$$\widehat{g}(\boldsymbol{\lambda},\omega) = \frac{1}{-\mathrm{i}(\omega+\mathrm{i}\lambda^2 D)}. \tag{9b}$$

Equation (9b) should be compared with Eq. (5). Whereas in Eq. (5) all frequencies (with the exception of the single point at $\omega = 0$) are large with respect to ϵ, the resonance-broadened result (9b) opens up a continuous band of frequencies that are small with respect to the diffusion rate. For those frequencies, it is not valid to treat D (or $\widetilde{\boldsymbol{a}}$) as small and endeavor to calculate its effect *via* regular perturbation theory. Such an attempt leads to singularities that grow progressively worse with order; thus, with $\eta \doteq \lambda^2 D$ denoting the resonance broadening, one has formally

$$\frac{1}{-\mathrm{i}(\omega+\mathrm{i}\eta)} = \frac{1}{-\mathrm{i}\omega}\left[1-\mathrm{i}\frac{\eta}{\omega}-\left(\frac{\eta}{\omega}\right)^2+\cdots\right]. \tag{10}$$

Only the sum of the series through all orders reconstructs the well-behaved function (9b).

2.4. Renormalization

It is undesirable to be forced to sum such series explicitly for several reasons: (i) in realistic situations, the values of most of the terms in the infinite series are not available in any practical way; (ii) an asymptotic series, even through all orders, may not represent the function from which it is derived [54]. (The classic example is $\mathrm{e}^{-1/\epsilon} \sim 0 + 0\epsilon + 0\epsilon^2 + \cdots$.) It is better if one is never required to calculate the series in the first place. Very loosely, any technique that effects or, better, bypasses altogether the summation of an infinite series is called *renormalization.* The term arises from quantum field theory, where one is concerned with determining the physically measurable (observable) mass and charge of a particle.

Renormalization is a vast subject [55] that certainly cannot be delved deeply into here. But it is useful to mention some of the consequences, which can be surprising. In particular, renormalization can lead to *anomalous scalings* that cannot be predicted by naive dimensional analysis [56]. A model of random passive advection illustrates the basic point. Consider

$$\partial_t\widetilde{\psi}(\boldsymbol{x},t) + \widetilde{\boldsymbol{V}}(t)\cdot\boldsymbol{\nabla}\widetilde{\psi} = 0. \tag{11}$$

Here $\widetilde{\boldsymbol{V}}$ is a statistically specified random velocity. In reality, $\widetilde{\boldsymbol{V}}$ would depend on both $\boldsymbol{x}$ and t, but here let us ignore its $\boldsymbol{x}$ dependence for simplicity. Then the spatial Fourier transform of Eq. (11) is

$$\partial_t\widetilde{\psi}_{\boldsymbol{k}}(t) + \mathrm{i}\widetilde{\Omega}(t)\widetilde{\psi}_{\boldsymbol{k}} \doteq 0, \tag{12}$$

which introduces the random advection frequency $\widetilde{\Omega}(t) \doteq \boldsymbol{k}\cdot\widetilde{\boldsymbol{V}}(t)$. (This frequency also plays a role in the later considerations of zonal flows in Sec. 3.2.) For even greater ease of analysis, let us ignore the time dependence as well, i.e., $\widetilde{\Omega}(t) \to \widetilde{\Omega}$, a Gaussian random number with $\langle\widetilde{\Omega}\rangle = 0$

and $\langle\widetilde{\Omega}^2\rangle = \beta^2$. Thus, we have arrived at the stochastic oscillator equation (1) (repeated here for convenience),

$$\partial_t\widetilde{\psi} = -\mathrm{i}\widetilde{\Omega}\widetilde{\psi}, \tag{13}$$

but now with a physical understanding of $\widetilde{\Omega}$ as a measure of nonlinear advection.

Although Eq. (13) is nonlinear in stochastic quantities, it is dynamically linear (as emphasized by Kraichnan [18]); thus the exact solution is trivially $\widetilde{\psi}(t) = \exp(-\mathrm{i}\widetilde{\Omega}t)\widetilde{\psi}_0$ or $\widehat{\widetilde{\psi}}(\omega) = 2\pi\widetilde{\psi}_0\delta(\omega-\widetilde{\Omega})$. (I shall assume that $\widetilde{\psi}_0$ is a Gaussianly distributed random variable statistically independent of $\widetilde{\boldsymbol{V}}$.) Either of the t- or ω-dependent forms demonstrates conservative evolution of any single realization. Because the Gaussian PDF of $\widetilde{\Omega}$ is largest (and smooth) at $\widetilde{\Omega}=0$, the form of $\widehat{\widetilde{\psi}}(\omega)$ indicates a broad resonance at $\omega=0$; this is characteristic of strong turbulence. Note that although each realization is oscillatory, statistical averaging leads *via* phase mixing to a decaying correlation function:

$$C(t,t') = C(\tau) \doteq \langle\delta\psi(t)\delta\psi^*(t')\rangle = \exp(-\tfrac{1}{2}\beta^2\tau^2)\langle|\delta\psi_0|^2\rangle. \tag{14}$$

This phase-mixing-induced decay (closely related to Landau damping) is analogous to the effects of the "nonlinear scrambling" that occurs in dynamically nonlinear systems with sufficiently strong mixing properties [32].

Although in the present case one knows the exact solution, one does not have that luxury for realistic nonlinear PDEs, so one is led to consider approximate *equations* for relevant observables. In doing so, it turns out to be very useful to introduce the concept of a mean Green's function. Thus, consider

$$\partial_t\widetilde{\psi}(t) + \mathrm{i}\widetilde{\Omega}\widetilde{\psi} = \widehat{\eta}(t), \tag{15}$$

where a statistically sharp external source $\widehat{\eta}$ has been added.[4] The average response of the oscillator is characterized by the *mean Green's function* or *infinitesimal response function*[5]

$$R(t;t') \doteq \left\langle \frac{\delta\psi(t)}{\delta\widehat{\eta}(t')} \right\rangle\bigg|_{\widehat{\eta}=0}. \tag{16}$$

In general, one can postulate for R the formally closed *Dyson equation*

$$\partial_t R(t;t') + \int_{t'}^{t} d\bar{t}\,\Sigma(t;\bar{t})R(\bar{t};t') = \delta(t-t'). \tag{17}$$

Then the goal becomes the calculation of the *memory function* Σ.

Equation (17) is nonlocal in time, which is to be expected in any system that possesses a nonzero autocorrelation time. But if one wants to just capture the time scale, one may less ambitiously characterize the oscillator by the Markovian equation

$$\partial_\tau R(\tau) + \eta^{\mathrm{nl}}R = \delta(\tau), \tag{18}$$

where $\eta^{\mathrm{nl}} \doteq \int_0^\infty d\overline{\tau}\,\Sigma(\overline{\tau})$, and attempt to calculate η^{nl}.[6] Let us begin naively and try to do that perturbatively. Now at lowest order (quasilinear level), it is easy to find that

$$\Sigma^{\mathrm{QL}}(\tau) = \beta^2 R^{(0)}(\tau), \tag{19}$$

[4] Do not confuse the external source $\widehat{\eta}$ with the resonance broadening η. Unfortunately, there are not enough letters in the Greek alphabet. The notation $\widehat{\eta}$ is used for consistency with Ref. 31. Here the circumflex does not indicate a Fourier transform, but rather a particular kind of source.

[5] Setting $\widehat{\eta}$ to 0 after the functional differentiation is unncessary in this dynamically linear problem, but is required more generally.

[6] The superscript nl stands for nonlinear and serves to distinguish η^{nl} from linear damping that will be added later.

where $R^{(0)}(\tau) = H(\tau)$ is Green's function for the linear dynamics. Thus the quasilinear approximation is causal but otherwise constant. This unfortunately leads to a severe pathology for the predicted damping:

$$\eta^{\mathrm{QL}} = \int_0^\infty d\overline{\tau}\, \Sigma^{\mathrm{QL}}(\overline{\tau}) = \int_0^\infty d\overline{\tau}\, \beta^2 \underbrace{R^{(0)}(\overline{\tau})}_{1} = \begin{cases} 0 & (\beta = 0) \\ \infty & (\beta \neq 0), \end{cases} \tag{20}$$

which is a discontinuous and singular function of β. Such discontinuities and infinities are always signatures of trouble with the underlying approximations. (In this case, the difficulty is that $R^{(0)}$ does not decay in time.) The infinity arising at lowest order precludes further perturbation-theoretic development, since all higher-order terms would be built on the infinite lowest-order result and would have indeterminate value.

To circumvent this problem, one can regularize the infinity by adding some linear damping ν to Eq. (13):

$$\partial_t \widetilde{\psi} + \nu \widetilde{\psi} = -\mathrm{i}\widetilde{\Omega}\widetilde{\psi}. \tag{21}$$

Now the linear Green's function is $R^{(0)}(\tau) = H(\tau)\mathrm{e}^{-\nu\tau}$, so

$$\eta^{\mathrm{QL}} = \int_0^\infty d\overline{\tau}\, \Sigma^{\mathrm{QL}}(\overline{\tau}) = \beta^2/\nu, \tag{22}$$

which happily is a continuous function of β. With perturbation theory regularized, one can now proceed order by order and eventually sum through all orders. Although that process is still complicated [18], one crucial part of the result can be shown to be the replacement of $R^{(0)}$ by R, i.e.,[7]

$$\Sigma^{\mathrm{nl}}(\tau) \approx \beta^2 R(\tau). \tag{23}$$

Although R is not yet known, it has been assumed to obey the Markovian dynamics $\partial_t R + \nu R + \eta^{\mathrm{nl}} R = \delta(t - t')$, so one has the formal solution

$$R(\tau) = H(\tau)\exp[-(\nu + \eta^{\mathrm{nl}})\tau]. \tag{24}$$

Upon calculating the total damping rate by integrating Eq. (23), one obtains the self-consistent equation

$$\eta^{\mathrm{nl}} = \frac{\beta^2}{\nu + \eta^{\mathrm{nl}}}, \tag{25}$$

which is a nonlinear algebraic equation that can be solved for η^{nl}. Although that can obviously be done exactly, it is more instructive to note that the equation remains well behaved in the limit $\nu \to 0$ (recall that ν was inserted only as an artifice to regularize an infinity), in which case one easily finds $\eta^{\mathrm{nl}} \to \beta$. It is remarkable that this damping rate is finite in the limit $\nu \to 0$ although it was infinite in the lowest-order theory with $\nu = 0$.

It is important to note that the renormalized solution exhibits *anomalous scaling* in its dependence on the advection level β:

$$\eta^{\mathrm{nl}} = \begin{cases} \beta^2/\nu & \text{(quasilinear theory),} \\ \beta & \text{(renormalized).} \end{cases} \tag{26}$$

Furthermore, the correct exponent of β, namely 1, cannot be predicted from dimensional analysis. Given that the problem is characterized by the two frequencies ν and β, dimensional analysis predicts merely that $\eta^{\mathrm{nl}} \propto \nu^a \beta^b$ with $a + b = 1$. Only the renormalization, which

[7] This is the direct-interaction approximation for the stochastic oscillator [18].

Journal of Physics: Conference Series **169** (2009) 012005 doi:10.1088/1742-6596/169/1/012005

formally involves consideration of an infinite number of terms, is capable of selecting the proper exponent.[8]

A fundamental advance in the theory of renormalization was made by MSR [31], who generalized to classical statistical dynamics important generating-functional methods of quantum field theory (the Schwinger formalism [57, 58]). The basic observation is that all statistical observables depend functionally on external sources. Even in cases in which such sources are absent in the true physical situation (consider the Vlasov equation, for example), it is useful to introduce them anyway because (i) the cumulants can be shown to be functional derivatives of a certain generating functional with respect to the sources; and (ii) the induced dependencies on the sources can (often) be used to deduce, *via* a functional chain rule, functional relations among the observables themselves. Those persist even in the limit of vanishing sources. One is led (see Sec. 2.6.2) to a system of coupled Dyson equations for C and R (plus the exact equation for the mean field). Approximation comes in the truncation of functional equations for the constituents of $\Sigma^{\rm nl}$. At lowest order, Kraichnan's DIA emerges. Of course, understanding the nature of the approximation is quite nontrivial. Martin has stressed that the MSR formalism is "not a panacea" [59]. However, the basic point remains that by proceeding in that way one is led to a self-consistent expression for the memory function. Note that giving any nonzero value at all to $\Sigma^{\rm nl}$ broadens the resonance predicted by the zeroth-order Green function and thus subsumes the effects of an infinite number of terms. That is, instead of writing $R = [(R^{(0)})^{-1} + \Sigma^{\rm nl}]^{-1}$, which cannot be approximated as $R^{(0)}$ near $(R^{(0)})^{-1} = 0$, one considers $R^{-1} = (R^{(0)})^{-1} + \Sigma^{\rm nl}$ and approximates $\Sigma^{\rm nl}$. This philosophy was well explained in the clearly written and highly recommended introduction to the original paper by MSR [31].

2.5. The role of dissipation

In the previous example, dissipation was introduced only as a way of regularizing a divergent perturbation theory. But in classical physics dissipation is virtually always present on physical grounds, the basic paradigm being the Navier–Stokes equation, which contains the viscosity (momentum-diffusion) term $\mu\nabla^2$. In such situations the statistical dynamics contain a built-in (generally short-wavelength) cutoff. Thus the MSR/Dyson equations make no reference to a cutoff, and the focus of classical renormalization shifts to effects of statistical self-consistency.

It is important to distinguish dissipation of the primitive amplitude from dissipative effects that emerge after statistical averaging. The diffusion equation (7) is dissipative, yet it arises from a conservative amplitude equation. The same is true for the $\Sigma^{\rm nl}$ calculated for the renormalized $\nu = 0$ stochastic oscillator.

Unfortunately, dissipation in the primitive amplitude equation is incompatible with many of the elegant tools derived for conservative systems, including Hamiltonian formulations and construction of equilibrium ensembles using notions of Gibbsian statistical mechanics [60, and references therein]. With characteristic insight, Prof. Kaufman made a start on this problem with his 1984 introduction of ***dissipative brackets*** [61] (see also the related work of Morrison in Ref. 62 and these Proceedings.) That interesting formalism is well suited to discussion of Markovian kinetic equations such as the Fokker–Planck equation with the Landau collision operator. However, it has not as yet been shown to coexist gracefully with the more fundamental non-Markovian statistical formalisms. Thus, the techniques that so far have proven to be most useful for the general turbulence problem, which is typically forced and dissipative, have a different flavor than most of the basic tools developed by Kaufman and his collaborators.

[8] One is led immediately to the proper answer by arguing that the ultimate statistics should not depend on ν at all. (Indeed, ν was absent from the original equation.) But in the Markovian perturbation theory for the example, it is impossible to ignore ν and deal with finite terms. A belief that the anomalous exponent of 1 is trivial for this problem amounts to an acceptance of the need for renormalization, which is exactly the point.

Journal of Physics: Conference Series **169** (2009) 012005 doi:10.1088/1742-6596/169/1/012005

Nevertheless, some of those ideas and results are hiding in the general equations of classical statistical dynamics [31], as we will see in the next section.

2.6. Renormalized spectral balance, ponderomotive effects, and the search for a renormalized K–χ theorem

In this section I discuss how ponderomotive effects are embedded in the general renormalized equations of plasma turbulence, and I present some initial steps toward the possible derivation of a so-called *renormalized K–χ theorem.* In the interest of full disclosure, I must state at the outset that, although various nontrivial details have been worked out, true understanding has not yet been achieved and it is far from obvious that a renormalized K–χ theorem even exists.

2.6.1. The classical K–χ theorem. At the time I visited Berkeley in the middle 1970's, one of Allan's many interests was ponderomotive force, especially its systematic derivation from Hamiltonian- and Lie-theoretic perturbation methods. Dewar had already published his fundamental paper on "Oscillation center quasilinear theory" in 1973 [4] and had followed that up with his ambitious 1976 paper on renormalized perturbation theory [5], in which he gave an important operator generalization of Deprit's Lie perturbation method [63]. Shortly thereafter, Cary and Kaufman published their work on "Ponderomotive force and linear susceptibility in Vlasov plasmas" [6], in which the so-called *K–χ theorem* was first stated; a concise proof was soon published by Johnston and Kaufman [39]. That theorem predicts a specific and simple relation between (i) the second-order ponderomotive Hamiltonian $K^{(2)}$ for a particle in a slightly inhomogeneous but rapidly oscillating wave field, and (ii) the (linear, or lowest-order) susceptibility $\chi^{(0)}$. The theorem can be stated quite generally, as was elegantly demonstrated later by Kaufman [64], but I shall just discuss a simple special case for illustration. Consider a temporally sinusoidal electric field $\boldsymbol{E}_\omega(\boldsymbol{x})$. With $(\boldsymbol{X}, \boldsymbol{P}) \equiv \Gamma$ being the oscillation-center position and momentum, Cary and Kaufman quote the theorem in the form

$$K^{(2)}(\Gamma) = -\frac{1}{4\pi}\int d\boldsymbol{x}\, d\boldsymbol{x}'\, \boldsymbol{E}^*_\omega(\boldsymbol{x}) \cdot \frac{\delta\boldsymbol{\chi}^{(0)}(\boldsymbol{x}, \boldsymbol{x}')}{\delta f(\Gamma)} \cdot \boldsymbol{E}_\omega(\boldsymbol{x}'), \tag{27}$$

where f is the background distribution function. For the special case of electrons with charge e and mass m in an electrostatic wave of wave number $\boldsymbol{k}$ with weakly inhomogeneous field intensity $\mathcal{E}_{\boldsymbol{k},\omega}(\boldsymbol{X}) \doteq |E_{\boldsymbol{k},\omega}(\boldsymbol{X})|^2$, the result is

$$K^{(2)}(\Gamma) = \left(\frac{e^2}{m}\right)\frac{\mathcal{E}_{\boldsymbol{k},\omega}(\boldsymbol{X})}{[\omega - \boldsymbol{k}\cdot\boldsymbol{v}(\boldsymbol{P})]^2}. \tag{28}$$

The Hamiltonian equations of motion

$$\frac{\partial K}{\partial \boldsymbol{P}} = \dot{\boldsymbol{X}}, \tag{29a}$$

$$-\frac{\partial K}{\partial \boldsymbol{X}} = \dot{\boldsymbol{P}} \equiv \boldsymbol{F} \tag{29b}$$

thus identify the ponderomotive force on the oscillation center as $\boldsymbol{F} = -\boldsymbol{\nabla} K$. Here the gradient operates on the weak inhomogeneity of the field intensity $\mathcal{E}$.

It is interesting to inquire how this result relates to statistical turbulence theory. That there should be a connection is obvious on physical grounds: ponderomotive force is a real physical effect that should not disappear when a single coherent wave is generalized to an inhomogeneous ensemble of turbulent waves. More abstractly, one is motivated by the observation that the *forms* of the statistical turbulence equations do not depend on the precise way in which the

Journal of Physics: Conference Series **169** (2009) 012005 doi:10.1088/1742-6596/169/1/012005

averaging procedure $\langle \ldots \rangle$ is implemented. Since the oscillation-center transformation, which systematically removes rapid oscillations, is a kind of averaging operation, one may expect that ponderomotive effects are included in the general statistical theory, and this will be seen to be true. On the other hand, no *statistical* averaging is done in the conventional derivations of either the ponderomotive force or the linear susceptibility, and that is a great simplification. In general, the presence of nontrivial statistical correlations at higher orders greatly complicates the general discussion, and a significant lack of complete understanding of those correlations turns out to be the sticking point in the derivation of a renormalized K–χ theorem.

2.6.2. General structure of the statistical theory of Vlasov turbulence; the renormalized dielectric function. Unfortunately, the general statistical theory of the Vlasov equation is somewhat tedious. It must, of course, be compatible with weak-turbulence theory [12] (which can be derived by a regular-perturbation-theoretic treatment of the Vlasov equation [14]), including at lowest nonlinear order wave–wave–wave and wave–wave–particle interactions. In addition, it must be renormalized in order to deal with resonance-broadening effects. Most of the details [28, 40] cannot be discussed here.

However, the general structure of the Dyson equations that couple Vlasov response and correlations is reasonably straightforward and instructive to consider. I shall first discuss the structure for a generic quadratically nonlinear equation of the form

$$\partial_t \psi(1) = L(1,2)\psi(2) + \frac{1}{2}M(1,2,3)\psi(2)\psi(3), \tag{30}$$

where a numerical argument refers to a set of dependent variables such as $1 \equiv \{\boldsymbol{x}_1, \boldsymbol{v}_1, s_1, t_1\}$ (s being a species label); summation/integration over repeated arguments is assumed. The Vlasov equation ($\psi \to f$), which possesses the quadratic integro-differential nonlinearity $(q/m)\boldsymbol{E} \cdot \partial_{\boldsymbol{v}} f$, can be written in this form by an appropriate choice of mode-coupling coefficient M. To do so, I consider electrostatics for simplicity, write $\boldsymbol{E}(1) = \widehat{\boldsymbol{E}}(1,2)f(2)$, where $\widehat{\boldsymbol{E}}$ is the linear operator that inverts Poisson's equation,[9] and adopt the shorthand notation $\boldsymbol{\partial} \doteq (q/m)\partial_{\boldsymbol{v}}$. Then

$$M_{\mathrm{Vlasov}}(1,2,3) = \widehat{\boldsymbol{E}}(1,2) \cdot \boldsymbol{\partial}_1 \delta(1-3) + (2 \Leftrightarrow 3). \tag{31}$$

(Symmetrization is possible here because the electric field is allowed to self-consistently respond to changes in the distribution function. It is not permissible in a passive problem in which the field is specified; see the calculation below.)

Equation (30) generates the exact equation for the mean field

$$\partial_t \langle\psi\rangle(1) = L(1,2)\langle\psi\rangle(2) + \frac{1}{2}M(1,2,3)\langle\psi\rangle(2)\langle\psi\rangle(3) + \frac{1}{2}M(1,2,3)C(2,3), \tag{32}$$

where $C(1,1') \doteq \langle\delta\psi(1)\delta\psi(1')\rangle$ and $\delta\psi \doteq \psi - \langle\psi\rangle$. From the exact equation for the fluctuation,

$$\partial_t \delta\psi = L\delta\psi + M\langle\psi\rangle\delta\psi + \frac{1}{2}M(\delta\psi\,\delta\psi - C), \tag{33}$$

one sees that the linear operator is modified by a mean-field term: $L \to \widehat{L} \doteq L + M\langle\psi\rangle$. [For the self-consistent Vlasov equation, the effective linear operator is $\widehat{L} = -(\boldsymbol{v} \cdot \boldsymbol{\nabla} + \boldsymbol{E} \cdot \boldsymbol{\partial} + \boldsymbol{\partial} f \cdot \widehat{\boldsymbol{E}})$.] Then the Dyson equation for the two-point response function $R(1;1')$ has the form

$$\partial_t R(1;1') - \widehat{L}R + \Sigma(1;\overline{1})R(\overline{1};1') = \delta(1-1'). \tag{34}$$

[9] Specifically, $\widehat{\boldsymbol{E}}_{\boldsymbol{k},\omega}(s,\boldsymbol{v};\overline{s},\overline{\boldsymbol{v}}) = -(4\pi\mathrm{i}\boldsymbol{k}/k^2)(\overline{n}q)_{\overline{s}}$, where $\overline{n}$ is the mean density.

Σ is a generalized "turbulent collision operator"; it describes the effects of resonance broadening, self-consistent polarization processes, (part of) nonlinear mode coupling, *etc.* Its precise form depends on the particular statistical closure, such as the DIA; it is a nonlinear functional of C and R itself.[10] It is frequently useful to think of Σ as a generalized Fokker–Planck operator that contains turbulent diffusion and drag terms [28, 65].

Response R is self-consistently coupled through Σ to fluctuations C. The Dyson equation for C is

$$\partial_t C(1,1') - \widehat{L}C + \Sigma(1;\overline{1})C(\overline{1},1') = F(1,\overline{1})R(1';\overline{1}). \tag{35}$$

Here a new function F (also functionally dependent on C and R) has been introduced. Upon observing that the operator on the left-hand side of Eq. (35) is Green's function for Eq. (34), namely R^{-1}, the C equation can be written in the symmetrical form

$$C = RFR^T, \tag{36}$$

where T denotes transpose. This is one form of the *spectral balance equation* for turbulence. The positive-definite function F may be thought of as the covariance of a certain internal *incoherent noise* arising from the nonlinearity. That forcing competes with *coherent nonlinear damping* (the Σ term in the R equation); the resulting balance sets the fluctuation level C. In addition to the detailed review of spectral balance in Ref. 28, further pedagogical words about the spectral balance equation can be found in the author's recent review article on gyrokinetic noise [45].[11]

For nonlinear fluid equations (e.g., the Navier–Stokes equation or the Hasegawa–Mima equation), the Dyson equations can be studied as is. The most sophisticated yet (sometimes) workable approximation is the DIA; Markovian closures such as the EDQNM exist as well. Application to the Vlasov equation requires further discussion because of the multi-level nature of the physics: the fundamental random variable (the Klimontovich distribution function) lives in a 6D phase space, but the electric field, which defines the nonlinearity, lives in the 3D $\boldsymbol{x}$ space. This leads to a decomposition of the damping operator Σ into two parts: one that applies the $\widehat{\boldsymbol{E}}$ operator; and one that operates directly on a phase-space function. I shall use the notation $\Sigma R = \Sigma^{(g)}R + (\boldsymbol{\partial}\Delta\overline{f})\cdot\widehat{\boldsymbol{E}}R$. (The significance of the g superscript will become apparent momentarily.) The symbolic notation $\boldsymbol{\partial}\Delta\overline{f}$ denotes a certain combination of nonlinear terms [28] whose details will not be required here. The notation has been chosen[12] such that this latter combination formally adds to the mean-field contribution $\boldsymbol{\partial}f\cdot\widehat{\boldsymbol{E}}R$; I write $\boldsymbol{\partial}\overline{f} \doteq \boldsymbol{\partial}f + \boldsymbol{\partial}\Delta\overline{f}$. Thus, the R equation for electrostatic Vlasov turbulence becomes

$$(\partial_t + \boldsymbol{v}\cdot\boldsymbol{\nabla} + \Sigma^{(g)})R(\boldsymbol{x},\boldsymbol{v},s,t;\boldsymbol{x}',\boldsymbol{v}',s',t') + \boldsymbol{\partial}\overline{f}\cdot\widehat{\boldsymbol{E}}R = \delta(t-t')\delta(\boldsymbol{x}-\boldsymbol{x}')\delta(\boldsymbol{v}-\boldsymbol{v}')\delta_{s,s'}. \tag{37}$$

The signficance of this form is that it closely resembles the linearized Vlasov equation

$$(\partial_t + \boldsymbol{v}\cdot\boldsymbol{\nabla})\delta f + \boldsymbol{\partial}f^{(0)}\cdot\delta\boldsymbol{E} = 0, \tag{38}$$

the differences being the addition of a resonance-broadening term $\Sigma^{(g)}$ and the generalization $f^{(0)} \to \overline{f}$. This strongly suggests that a renormalized dielectric $\mathcal{D}$ can be defined by following the same operational procedure used for calculating the lowest-order dielectric from Eq. (38). If

[10] In the general MSR formalism, a third (vertex) function Γ (actually a matrix with three independent components) is introduced and closures are characterized by the approximations made to the functional equation for Γ. It is not necessary to delve into the details [28] here.

[11] A forthcoming paper [66] (see Sec. 2.7 of the present paper) contains a useful appendix on spectral balance for a particular model of passive advection.

[12] The notation disguises the fact that $\boldsymbol{\partial}\Delta\overline{f}$ is not actually the velocity gradient of a scalar, but rather is the divergence of a (wave-number-dependent) tensor.

g is the renormalized particle propagator, defined by omitting from Eq. (37) the self-consistent response term $\boldsymbol{\partial}\overline{f}\cdot\widehat{\boldsymbol{E}}R$, then one has schematically

$$g^{-1}R+\boldsymbol{\partial}\overline{f}\cdot\widehat{\boldsymbol{E}}R=\mathsf{I}. \tag{39}$$

(I stands for the appropriate delta functions.) The solution of this equation is

$$\widehat{\boldsymbol{E}}R = \mathcal{D}^{-1}\widehat{\boldsymbol{E}}g, \tag{40a}$$
$$R = g-g\,\boldsymbol{\partial}\overline{f}\mathcal{D}^{-1}\cdot\widehat{\boldsymbol{E}}g, \tag{40b}$$

where the *renormalized dielectric function* is

$$\mathcal{D}\doteq 1+\widehat{\boldsymbol{E}}g\cdot\boldsymbol{\partial}\overline{f}. \tag{41}$$

This has precisely the form of the well-known lowest-order (electrostatic) dielectric

$$\mathcal{D}^{(0)}(\boldsymbol{k},\omega)=1+\sum_s\frac{\omega_{ps}^2}{k^2}\int d\boldsymbol{v}\,\frac{\boldsymbol{k}\cdot\partial_{\boldsymbol{v}}f^{(0)}}{\omega-\boldsymbol{k}\cdot\boldsymbol{v}+\mathrm{i}\epsilon} \tag{42}$$

with the replacements $g^{(0)}\to g$ and $\boldsymbol{\partial}f^{(0)}\to\boldsymbol{\partial}\overline{f}$. Here $g^{(0)}$ is the linear particle propagator

$$g^{(0)}(\boldsymbol{x},\boldsymbol{v},s,t;\boldsymbol{x}',\boldsymbol{v}',s',t') = H(\tau)\delta(\boldsymbol{\rho}-\boldsymbol{v}\tau)\delta(\boldsymbol{v}-\boldsymbol{v}')\delta_{s,s'}, \tag{43a}$$
$$\widehat{g}^{(0)}_{\boldsymbol{k},\omega}(\boldsymbol{v},s;\boldsymbol{v}',s') = \frac{\delta(\boldsymbol{v}-\boldsymbol{v}')\delta_{s,s'}}{-\mathrm{i}(\omega-\boldsymbol{k}\cdot\boldsymbol{v}+\mathrm{i}\epsilon)}, \tag{43b}$$

where $\boldsymbol{\rho}\doteq\boldsymbol{x}-\boldsymbol{x}'$ and the Fourier transforms are with respect to $\boldsymbol{\rho}$ and τ.

Although this approach to the form of the nonlinear dielectric function is suggestive, it may not be apparent that the resulting form (41) is compatible with the conventional definition of a dielectric function that is made in linear response theory. The formal proof was first given by DuBois and Espedal [40]; an alternate derivation can be found in Ref. 28, Sec. 6.5.2, p. 165.

It is important to stress that the previous discussion applies to dynamically *self-consistent* problems in which the particles are themselves responsible for the fields: $\boldsymbol{E}=\widehat{\boldsymbol{E}}f$. One can also consider problems of *passive advection* (usually called *stochastic acceleration* [67] in the Vlasov context), in which the electric field is statistically specified. That has consequences for the forms of both the primitive amplitude equation and the renormalized Dyson equations. One is no longer allowed to write $\boldsymbol{E}=\widehat{\boldsymbol{E}}f$, since $\boldsymbol{E}$ is imposed. This implies that the symmetrized form (30) may not be used. Because self-consistent response is forbidden, various "polarization" effects [28, 65] are absent. Nevertheless, a nontrivial, renormalized particle propagator g does exist in the passive problem. One can think of g as the conditional PDF of finding a test particle, moving under the influence of a random electric field, at phase-space point $\{\boldsymbol{x},\boldsymbol{v}\}$ at time t, given that it was at $\{\boldsymbol{x}',\boldsymbol{v}'\}$ at time t'. In the next section I will show how to extract ponderomotive force from the Dyson equation for the passive g.

2.6.3. Ponderomotive force effects from the quasilinear theory of the weakly inhomogeneous particle propagator. The nonlinear dielectric might provide one piece of a renormalized K–χ theorem. Before attempting to draw any such connection, however, one must identify a statistical version of ponderomotive force. Because the self-consistent problem is quite complicated, I shall here consider stochastic acceleration of a test particle. (This is consistent with the fact that the original K–χ theorem assumed a given wave field.) Because such a test particle would respond to ponderomotive force, one should examine the nonlinear contributions to g in a search for a statistical expression of ponderomotive effects. That is, one must study $\Sigma^{(g)}$. To illustrate the relevant issues with a minimum of complexity, I shall proceed by using the direct-interaction approximation. Of course, this omits a host of statistical effects (contained in the vertex renormalizations that are ignored in the DIA) and therefore precludes a discussion of complete generality. I shall also restrict attention to a single species.

It can be shown that in the passive DIA one finds

$$\Sigma^{(g)}(1,\overline{1}) = -\boldsymbol{\partial}_1 \cdot g(1,2)\boldsymbol{\mathcal{E}}(2,1)\cdot\boldsymbol{\partial}_2\delta(2-\overline{1}), \tag{44}$$

where $\boldsymbol{\mathcal{E}}(1,1') \doteq \langle\delta\boldsymbol{E}(1)\delta\boldsymbol{E}(1')\rangle$. The presence of the two velocity derivatives identifies $\Sigma^{(g)}$ as a generalized diffusion operator. To get a feeling for its content, first assume that spatial homogeneity and temporal stationarity prevail (on the average), so that all two-point functions depend on only the differences of their spatial and temporal arguments. Those functions can then be represented in Fourier space; one finds[13]

$$\widehat{\Sigma}^{(g)}_{\boldsymbol{k},\omega}(\boldsymbol{v};\overline{\boldsymbol{v}}) = -\boldsymbol{\partial}_1 \cdot \int \frac{d\overline{\boldsymbol{p}}}{(2\pi)^3}\frac{d\omega_{\overline{\boldsymbol{p}}}}{2\pi}\, g_{\boldsymbol{k}-\overline{\boldsymbol{p}},\omega-\omega_{\overline{\boldsymbol{p}}}}(\boldsymbol{v}_1;\boldsymbol{v}_2)\boldsymbol{\mathcal{E}}_{\overline{\boldsymbol{p}},\omega_{\overline{\boldsymbol{p}}}}\cdot\boldsymbol{\partial}_2\delta(\boldsymbol{v}_2-\overline{\boldsymbol{v}}). \tag{45}$$

This can be reduced to the familiar quasilinear diffusion operator by (i) passing to the long-wavelength limit $|\boldsymbol{k}| \ll |\overline{\boldsymbol{p}}|$ and the low-frequency limit $\omega \ll \omega_{\overline{\boldsymbol{p}}}$; (ii) assuming that the imposed field spectrum is wave-like: $\boldsymbol{\mathcal{E}}_{\boldsymbol{k},\omega} = 2\pi\boldsymbol{\mathcal{E}}_{\boldsymbol{k}}\delta(\omega-\Omega_{\boldsymbol{k}})$, where $\Omega_{\boldsymbol{k}}$ is a specified frequency; and (iii) approximating g by the unperturbed propagator $g^{(0)}$ [which contains $\delta(\boldsymbol{v}_1-\boldsymbol{v}_2)$]. Then

$$\Sigma^{(g)}(\boldsymbol{v};\overline{\boldsymbol{v}}) \approx -\boldsymbol{\partial}_1\cdot\left(\int\frac{d\overline{\boldsymbol{p}}}{(2\pi)^3}\,\pi\delta(\Omega_{\overline{\boldsymbol{p}}}-\overline{\boldsymbol{p}}\cdot\boldsymbol{v})\boldsymbol{\mathcal{E}}_{\overline{\boldsymbol{p}}}\right)\cdot\boldsymbol{\partial}_1\delta(\boldsymbol{v}_1-\overline{\boldsymbol{v}}). \tag{46}$$

The parenthesized term is the well-known quasilinear diffusion tensor $\mathsf{D}(\boldsymbol{v})$ (the significance of which was elaborated so well by Kaufman [2]).

Of course, a statistically homogeneous formalism cannot be expected to describe ponderomotive force; one must allow for weak inhomogeneities. The most elegant treatment of that topic involves the Weyl calculus, reviewed by McDonald [7]. To lowest order, however, the formalism is well known [68] and can be worked out straightforwardly [69]. For simplicity, ignore time dependence temporarily and write an arbitrary two-point space-dependent function A in terms of sum and difference variables: $A(\boldsymbol{x},\boldsymbol{x}') = A(\boldsymbol{x}-\boldsymbol{x}' \mid \frac{1}{2}(\boldsymbol{x}+\boldsymbol{x}')) = A(\boldsymbol{\rho}\mid\boldsymbol{X})$; one may Fourier transform with respect to the difference argument. To first order in a weak inhomogeneity, the operation $\int d\overline{1}\,A(1,\overline{1})B(\overline{1},1') \equiv (A\star B)(1,1')$ can be written as

$$\widehat{(A\star B)}_{\boldsymbol{k}}(\boldsymbol{X}) \approx \widehat{A}_{\boldsymbol{k}}(\boldsymbol{X})\widehat{B}_{\boldsymbol{k}}(\boldsymbol{X}) + \frac{1}{2}\mathrm{i}\{\widehat{A},\widehat{B}\}, \tag{47}$$

where

$$\{\widehat{A},\widehat{B}\} \doteq \frac{\partial\widehat{A}}{\partial\boldsymbol{X}}\cdot\frac{\partial\widehat{B}}{\partial\boldsymbol{k}} - \frac{\partial\widehat{A}}{\partial\boldsymbol{k}}\cdot\frac{\partial\widehat{B}}{\partial\boldsymbol{X}} = \frac{\partial}{\partial\boldsymbol{X}}\cdot\left(\widehat{A}\frac{\partial\widehat{B}}{\partial\boldsymbol{k}}\right) - \frac{\partial}{\partial\boldsymbol{k}}\cdot\left(\widehat{A}\frac{\partial\widehat{B}}{\partial\boldsymbol{X}}\right). \tag{48}$$

This is promising because gradients with respect to the weak inhomogeneity have appeared. When time dependence is included, the representation generalizes to $A(\boldsymbol{x},t,\boldsymbol{x}',t') = A(\boldsymbol{x}-\boldsymbol{x}',t-t' \mid \frac{1}{2}(\boldsymbol{x}+\boldsymbol{x}'),\frac{1}{2}(t+t')) \equiv A(\boldsymbol{\rho},\tau\mid\boldsymbol{X},T)$. I shall consider only steady-state turbulence,

[13] One is presented here with a classic collision of notation. A common convention in homogeneous statistical theory theory is to consider wave-number triads called $(\boldsymbol{k},\boldsymbol{p},\boldsymbol{q})$ with $\boldsymbol{k}+\boldsymbol{p}+\boldsymbol{q}=\boldsymbol{0}$. Unfortunately, use of $\boldsymbol{p}$ here would preclude its use as a canonical momentum, to which I will also need to refer later. Therefore, in the wave-number context I shall use $\overline{\boldsymbol{p}}$ instead of $\boldsymbol{p}$; hopefully that will not be confused with a canonically transformed momentum.

Journal of Physics: Conference Series **169** (2009) 012005 doi:10.1088/1742-6596/169/1/012005

so no dependence with respect to T appears and temporal corrections to the Poisson bracket vanish. The formalism therefore remains correct if all AB products in Eqs. (47) and (48) are generalized to include a convolution in time difference τ (or, of course, simple products of $\widehat{A}_{\boldsymbol{k},\omega}$ and $\widehat{B}_{\boldsymbol{k},\omega}$). Thus, consider

$$\partial_\tau \widehat{g}_{\boldsymbol{k}}(\boldsymbol{X},\boldsymbol{v},\tau;\boldsymbol{v}') + \mathrm{i}\boldsymbol{k}\cdot\boldsymbol{v}\widehat{g}_{\boldsymbol{k}} + \boldsymbol{v}\cdot\boldsymbol{\nabla}\widehat{g}_{\boldsymbol{k}} + (\widehat{\Sigma^{(g)}\star g})_{\boldsymbol{k}}(\boldsymbol{X},\boldsymbol{v},\tau;\boldsymbol{v}') = \delta(\tau)\delta(\boldsymbol{v}-\boldsymbol{v}'), \tag{49}$$

where

$$\begin{aligned}(\widehat{\Sigma^{(g)}\star g})_{\boldsymbol{k}}(\boldsymbol{X},\boldsymbol{v},\tau;\boldsymbol{v}') \approx \int_0^\tau d\overline{\tau}\,\Big(&\widehat{\Sigma}^{(g)}_{\boldsymbol{k}}(\boldsymbol{X},\boldsymbol{v},\overline{\tau};\overline{\boldsymbol{v}})\widehat{g}_{\boldsymbol{k}}(\boldsymbol{X},\overline{\boldsymbol{v}},\tau-\overline{\tau};\boldsymbol{v}')\\ &+\frac{1}{2}\mathrm{i}\{\widehat{\Sigma}^{(g)}_{\boldsymbol{k}}(\boldsymbol{X},\boldsymbol{v},\overline{\tau};\overline{\boldsymbol{v}}),\widehat{g}_{\boldsymbol{k}}(\boldsymbol{X},\overline{\boldsymbol{v}},\tau-\overline{\tau};\boldsymbol{v}')\}\Big),\end{aligned} \tag{50}$$

and focus in particular on the Poisson-bracket terms. To isolate specific effects related to force, one may examine the time development of $\langle\boldsymbol{v}\rangle(\boldsymbol{\rho},\tau) \doteq \int d\boldsymbol{v}\int d\boldsymbol{X}\,\boldsymbol{v}g(\boldsymbol{\rho},\tau,\boldsymbol{v}\mid\boldsymbol{X},\boldsymbol{v}')$. In linear theory (i.e., with the substitution $g\to g^{(0)}$), one would find $\langle\boldsymbol{v}\rangle = H(\tau)\boldsymbol{v}'\delta(\boldsymbol{\rho}-\boldsymbol{v}'\tau)$, which describes free-streaming propagation of a localized pulse. In general, one has

$$\partial_\tau\widehat{\langle\boldsymbol{v}\rangle}_{\boldsymbol{k}} + \mathrm{i}\boldsymbol{k}\cdot\widehat{\mathsf{P}}_{\boldsymbol{k}} + \int d\boldsymbol{v}\int d\boldsymbol{X}\,(\widehat{\Sigma^{(g)}\star g})_{\boldsymbol{k}}(\boldsymbol{X},\boldsymbol{v},\tau;\boldsymbol{v}') = \delta(\tau)\delta(\boldsymbol{v}-\boldsymbol{v}'), \tag{51}$$

where $\mathsf{P}\doteq\langle\boldsymbol{v}\,\boldsymbol{v}\rangle$. The last term on the left-hand side can be interpreted as $-\boldsymbol{\mathcal{F}}/m$, where $\boldsymbol{\mathcal{F}}$ is the averaged effect of nonlinearity-induced forces. Eventually the limit $k\to 0$ will be taken, so the $\mathrm{i}\boldsymbol{k}\cdot\mathsf{P}$ term will be irrelevant.

In interpreting Eq. (51), one must not forget that it describes the temporal development of the mean *particle* velocity. The statistical formalism is impartial; it will report correct statistical relations for any consistent set of independent variables in which the primitive amplitude dynamics are formulated. Since sophisticated variables such as oscillation-center coordinates and momenta are not known until the nonlinear dynamics have been examined, it is usually the original particle coordinates that are used.[14] However, in perturbation theory the particle velocity differs on the average from the oscillation-center velocity by a second-order correction. I will return to this point after working out the predictions of Eq. (51).

Before considering the most general renormalized case (in which one must deal with the fact that g is nonlocal in velocity space due to resonance broadening), let us again make assumptions (ii) (wave spectrum) and (iii) ($g\to g^{(0)}$) above. Also consider times long compared to a characteristic autocorrelation time of the turbulence, so that a Markovian approximation is valid. That is, upon writing out just the temporal arguments of the required convolution, one has

$$\int_0^\tau d\overline{\tau}\,\Sigma^{(g)}(\overline{\tau})g(\tau-\overline{\tau}) \approx \left(\int_0^\infty d\overline{\tau}\,\Sigma^{(g)}(\overline{\tau})\right)g(\tau) = \widehat{\Sigma}^{(g)}_{\omega=0}g(\tau). \tag{52}$$

Having performed the appropriate coarse-graining, one may obtain a definite answer by evaluating the nonlinear terms at $\tau=0$. The nonlinear contributions associated with weak inhomogeneity are then

$$\boldsymbol{\mathcal{F}}/m \approx -\frac{1}{2}\mathrm{i}\int d\boldsymbol{v}\int d\boldsymbol{X}\,\boldsymbol{v}\,\partial_{\boldsymbol{v}}\cdot\int\frac{d\overline{\boldsymbol{p}}}{(2\pi)^3}\frac{\partial}{\partial\boldsymbol{k}}\cdot[\widehat{g}^{(0)}_{\boldsymbol{k}-\overline{\boldsymbol{p}},-\Omega_{\overline{\boldsymbol{p}}}}(\boldsymbol{v})\boldsymbol{\mathcal{E}}_{\overline{\boldsymbol{p}}}\cdot\partial_{\boldsymbol{v}}\boldsymbol{\nabla}\widehat{g}_{\boldsymbol{k}}(\tau=0)], \tag{53}$$

where the second (conservative) form of Eq. (48) was used and the $\partial/\partial\boldsymbol{X}\equiv\boldsymbol{\nabla}$ was integrated away; $\widehat{g}^{(0)}_{\boldsymbol{k},\omega}(\boldsymbol{v})$ denotes Eq. (43b) *sans* the delta function in velocity. Both velocity derivatives

[14] But see Ref. 5 for some nontrivial generalizations.

can be integrated by parts, as can the ∇:

$$\mathcal{F}/m = \frac{1}{2}\mathrm{i}\int d\boldsymbol{v}\int d\boldsymbol{X}\int\frac{d\overline{\boldsymbol{p}}}{(2\pi)^3}\left(\nabla\cdot\frac{\partial}{\partial\boldsymbol{k}}[\partial_{\boldsymbol{v}}\widehat{g}^{(0)}_{\boldsymbol{k}-\overline{\boldsymbol{p}},-\Omega_{\overline{\boldsymbol{p}}}}(\boldsymbol{v})\cdot\boldsymbol{\mathcal{E}}_{\overline{\boldsymbol{p}}}]\right)\widehat{g}_{\boldsymbol{k}}(0). \tag{54}$$

(A term in $\partial_{\boldsymbol{k}}\widehat{g}_{\boldsymbol{k}}$ vanished at $\tau = 0$. The more general significance of that term will be disucssed elsewhere.) One has $\partial_{\boldsymbol{v}}\widehat{g}^{(0)}_{\boldsymbol{k}-\overline{\boldsymbol{p}},-\Omega_{\overline{\boldsymbol{p}}}} = (-\mathrm{i})^{-1}(\boldsymbol{k}-\overline{\boldsymbol{p}})/[-\Omega_{\overline{\boldsymbol{p}}} - (\boldsymbol{k}-\overline{\boldsymbol{p}})\cdot\boldsymbol{v}]^2$. Now the $\partial/\partial\boldsymbol{k}$ acts on two $\boldsymbol{k}$ dependencies: the $\boldsymbol{k}$ in the numerator, and the $(\boldsymbol{k}-\overline{\boldsymbol{p}})\cdot\boldsymbol{v}$ in the denominator; call those contributions $\mathcal{F}_a$ and $\mathcal{F}_b$. After evaluating them, take the limit $\boldsymbol{k}\to\boldsymbol{0}$ in the coefficient of $\widehat{g}_{\boldsymbol{k}}(0)$. (This could not be done before the differentiation with respect to $\boldsymbol{k}$ was performed.) One obtains $\mathcal{F}_a = \langle\boldsymbol{F}_a\rangle$, where the average is with respect to $\widehat{g}_{\boldsymbol{k}=\boldsymbol{0}}(\boldsymbol{v},\tau=0;\boldsymbol{v}')$ and

$$\boldsymbol{F}_a = -\nabla\left(\frac{1}{2}\frac{e^2}{m}\int\frac{d\overline{\boldsymbol{p}}}{(2\pi)^3}\frac{\nabla\boldsymbol{\mathcal{E}}_{\overline{\boldsymbol{p}}}}{(\Omega_{\overline{\boldsymbol{p}}}-\overline{\boldsymbol{p}}\cdot\boldsymbol{v})^2}\right). \tag{55}$$

For a wave field localized to a single wave number $\overline{\boldsymbol{p}}_0$ (and its negative), one obtains

$$\boldsymbol{F}_a = -\nabla K, \tag{56}$$

with

$$K \doteq \frac{e^2}{m}\frac{\boldsymbol{\mathcal{E}}_{\overline{\boldsymbol{p}}_0}}{(\Omega_{\overline{\boldsymbol{p}}_0}-\overline{\boldsymbol{p}}_0\cdot\boldsymbol{v})^2}. \tag{57}$$

This is the standard, well-known expression for the second-order ponderomotive force on the oscillation center; cf. Eq. (28).

Remaining is $\mathcal{F}_b$, which will be argued to be related to the second-order difference between the particle velocity and the oscillation-center velocity. To show that, I briefly review the standard development of the Lie perturbation theory for the oscillation-center canonical transformation. Recall [70, 71] that the transformation between old coordinates z and new coordinates $\overline{z}$ is generated by $\overline{z} = Tz$, where $T = T_0 + T_1 + T_2 + \cdots$. One has

$$T_0 = 1, \tag{58a}$$
$$T_1 = -L_1, \tag{58b}$$
$$T_2 = -\frac{1}{2}L_2 + \frac{1}{2}L_1^2, \tag{58c}$$

where the L's are defined in terms of the Poisson bracket with the generating function w: $L_n \doteq \{w_n,\cdot\}$. For example, the oscillation-center momentum $\boldsymbol{P}$ is obtained from the laboratory momentum $\boldsymbol{p}$ by

$$\boldsymbol{P} = \boldsymbol{p} + \boldsymbol{P}_1(z) + \boldsymbol{P}_2(z) + \cdots, \tag{59}$$

where, e.g., $\boldsymbol{P}_n \doteq T_n\boldsymbol{p}$. To avoid secularities and vitiation of the perturbative expansion, the w_n's are constructed to have zero average. Then

$$\langle\boldsymbol{P}\rangle = \boldsymbol{p} + \langle\boldsymbol{P}_2\rangle, \tag{60}$$

where

$$\langle\boldsymbol{P}_2\rangle = \frac{1}{2}\langle L_1^2\rangle\boldsymbol{p}. \tag{61}$$

After spatial integration by parts under the average, this can be rearranged into

$$\frac{1}{2}\langle L_1^2\rangle\boldsymbol{p} = \frac{1}{2}\frac{\partial}{\partial\boldsymbol{p}}\cdot\left\langle\frac{\partial w_1}{\partial\boldsymbol{x}}\frac{\partial w_1}{\partial\boldsymbol{x}}\right\rangle. \tag{62}$$

Journal of Physics: Conference Series **169** (2009) 012005 doi:10.1088/1742-6596/169/1/012005

The right-hand side of Eq. (62) can be evaluated explicitly by recalling that

$$(\partial_t + \boldsymbol{v}\cdot\boldsymbol{\nabla})w_1 = -q\varphi(\boldsymbol{x}, t). \tag{63}$$

It is left as an exercise to complete the calculations and show that the contribution $\langle\dot{\boldsymbol{P}}_2\rangle$ is precisely recovered by the $\boldsymbol{\mathcal{F}}_b$ term that was obtained above from the expansion in weak inhomogeneity.[15]

2.6.4. The possibility of a renormalized statistical K–χ theorem. The above manipulations can be repeated without the quasilinear approximation $g \to g^{(0)}$, and one is led thereby to DIA-renormalized expressions for the ponderomotive evolution of $\langle\boldsymbol{v}\rangle$ due to weak inhomogeneity.[16] The ease with which those generalizations follow from the renormalized theory makes a strong argument in favor of MSR-style renormalizations; no series need to be summed through all orders. The formulas naturally resolve the resonant-particle singularity. There are likely practical consequences to such renormalizations; however, those will not be pursued here.

Now consider the possibility of a renormalized K–χ theorem. Of course, no such theorem is required in order to evaluate the renormalized ponderomotive effects as above, but if such a theorem did exist it would imply interesting consistency constraints on approximations made to the dielectric function and the turbulent collision operator. However, one should probably not compare apples (passive advection) with oranges (self-consistent dynamics). I have shown how to work out renormalized ponderomotive effects from the $\Sigma^{(g)}$ appropriate for passive advection. The dielectric $\mathcal{D}$, however, is relevant to self-consistent problems. For passive problems, R reduces to g and $\mathcal{D} \to 1$. That is not helpful in formulating a K–χ theorem.

Nevertheless, suppose that one persists in attempting to derive passive ponderomotive force from the expression for the renormalized dielectric. Note that the construction $g\,\boldsymbol{\partial}$ appears in the general forms of both $\mathcal{D}$ and the renormalized $\Sigma^{(g)}$ operator. A reasonable first guess for a statistical K–χ theorem might then be[17] schematically

$$\boldsymbol{F} = -\boldsymbol{\nabla}(\widehat{K}\boldsymbol{\mathcal{E}}), \quad \widehat{K} = -\frac{\delta\mathcal{D}}{\delta\overline{f}}. \tag{64}$$

However, difficulties emerge immediately. The averaging required for the renormalized dielectric is over not the mean distribution function f but rather the effective (wave-number-dependent) distribution $\overline{f}$. It is unclear what functional differentiation with respect to $\overline{f}$ really means, and I have noted that the notation $\boldsymbol{\partial}\Delta\overline{f}$ is symbolic, being the velocity divergence of a tensor rather than literally the gradient of a scalar. Beyond the DIA, renormalized vertices that I have not discussed provide further essential complication. And if one contemplates deriving ponderomotive effects from the fully self-consistent problem, one must deal with the fact that a functional derivative with respect to the background distribution function (or any generalization thereof) would act on not just the explicit distribution in $\mathcal{D}$ but also the implicit $\Sigma^{(g)}$ in g.

The most immediate unresolved issue is that the fundamental significance of the turbulent correction $\Delta\overline{f}$ is not understood even after three decades of research on MSR-level renormalized plasma turbulence theory.[18] Its contributions to weak Vlasov turbulence theory are understood

[15] As a hint, note that $d\mathcal{E}(\boldsymbol{X})/dt = \boldsymbol{V}\cdot\boldsymbol{\nabla}\mathcal{E}$.

[16] One must deal with the fact that the resonance-broadened g is nonlocal in velocity space. Thus $\int d\boldsymbol{v}\,\boldsymbol{v}\,\partial_{\boldsymbol{v}} \cdot \widehat{g}_{\boldsymbol{k}-\overline{\boldsymbol{p}}}(\boldsymbol{v};\overline{\boldsymbol{v}})\cdot\partial_{\overline{\boldsymbol{v}}}\widehat{g}_{\boldsymbol{k}}(\overline{\boldsymbol{v}};\boldsymbol{v}') = \int d\boldsymbol{v}\,[\partial_{\overline{\boldsymbol{v}}}\widehat{g}_{\boldsymbol{k}-\overline{\boldsymbol{p}}}(\boldsymbol{v};\overline{\boldsymbol{v}})]\widehat{g}_{\boldsymbol{k}}(\overline{\boldsymbol{v}};\boldsymbol{v}')$; the last derivative is on the second velocity argument.

[17] Notice that the spatial gradient and the intermediate velocity derivative that are performed operate on the implicit $\Sigma^{(g)}$ that renormalizes g, not just merely the "obvious" dependencies.

[18] This remark is slightly misleading. To the best of my knowledge, no one has been continuously working on the physics of $\Delta\overline{f}$ since the initial flurry of activity in the late 1970's by such people as DuBois and Espedal [40] and Krommes [41, 72].

in the narrow sense that its relationships to specific terms in the third-order perturbative expansion of the Vlasov equation [28, 40, 72] have been identified. This means that in principle some of its connections to higher-order Hamiltonian/Lie perturbation theory are available; cf. the derivation of weak-turbulence effects by Johnston and Kaufman [39]. However, deep understanding of $\Delta\overline{f}$ as a "renormalization of the mean distribution function," as this author has loosely suggested [41], is lacking [40]. The origins of $\Delta\overline{f}$ lie in statistical correlations arising in the rigorous theory of infinitesimal response, which is used to define the renormalized dielectric function. As was described in Ref. 28, Sec. 6.5.2, p. 165, in order to determine the mean response to a small external potential one must evaluate $\langle\widetilde{R}\,\boldsymbol{\partial}\widetilde{f}\rangle$, where $\widetilde{R}$ is the random response function and $\widetilde{f}$ is the random distribution function. If that average would factor, one would have $\langle\widetilde{R}\,\boldsymbol{\partial}\widetilde{f}\rangle \to R\,\boldsymbol{\partial}f$, whereupon it is easy to show that one would obtain an expression for the renormalized dielectric $\mathcal{D}$ that has exactly the same form as the linear dielectric $\mathcal{D}^{(0)}$ except that $g^{(0)} \to g$ and $f_0 \to f$. Functional differentiation of that form with respect to f would at least be well defined. However, the statistical correlations between $\widetilde{R}$ and $\widetilde{f}$ are real and produce the $\Delta\overline{f}$ correction,[19] which is required in order to produce the correct renormalized normal modes of the turbulent plasma and for a proper description of nonlinear mode coupling.

From the linear response theory cited above, one can show that $\Delta\overline{f}$ is related to the response function $G_{+--}(1, 1', 1'')$, which describes the response at point 1 due to infinitesimal perturbations at both $1'$ and $1''$. What is missing is a proven simple connection between G_{+--} and the renormalized ponderomotive effects in which we are interested. If there is such a connection, it will likely not be deduced by detailed examination of renormalized closures such as the DIA, as I have attempted to do above, but rather from a new understanding of the general structure of the renormalized theory. Another way of saying this is that the original K–χ theorem follows immediately [39] from a relation between the second-order wave–particle interaction energy (which involves the first-order four-current and the first-order four-potential) and the second-order oscillation-center Hamiltonian. We have as yet no analog of that elegant construction in the renormalized theory. Note that the statistical correlations responsible for $\Delta\overline{f}$ do not exist between the first-order quantities.

The preceding discussion illustrates both the strengths and the weaknesses of statistical turbulence theory. On the positive side, formal renormalization *a la* MSR seems to effortlessly effect a statistical averaging that mirrors the Hamiltonian oscillation-center theory, and it deals with renormalized propagators from the outset. It also has great generality, dealing with interactions between scales of all sizes, including comparable ones. On the negative side, that same generality makes it difficult to see the forest for the trees, i.e., to distill specific physical effects from the complicated mixture of nonlinear statistical couplings. Furthermore, the Hamiltonian properties of the nonlinear interactions (if they exist[20]) do not appear to be explicitly used in any obvious way.

One is left with the uneasy feeling that some important conceptual understanding of the statistical theory remains to be achieved; certainly further work remains to be done before a renormalized K–χ theorem is proven, if indeed such a theorem exists at all. It is very likely that known results from Lie methodology and the other early work of Kaufman and his collaborators will provide important inspiration and guidance in any quest for deeper understanding. Unfortunately, the camps of statistical turbulence theory and of Lie perturbation theory have been only weakly coupled over the years. It would certainly be beneficial if that situation were corrected.

[19] Some terms in that correction exist even for passive advection.

[20] Note that the general MSR methodology does not require that the nonlinear term in the primitive amplitude equation be derivable from a Hamiltonian formalism.

Journal of Physics: Conference Series **169** (2009) 012005 doi:10.1088/1742-6596/169/1/012005

2.7. Application of renormalization to the theory of stochastic magnetic fields

A problem of considerable interest to magnetic fusion is the role of stochastic magnetic fields. Broken magnetic flux surfaces can lead to enhanced transport or even disruption of the discharge. The subject has a long history that cannot be reviewed here; see Ref. 73 for some of the early theoretical references.

A very recent calculation[21] involving stochastic magnetic fields [66] marries a variety of the analytical tools for turbulence with the very practical problem of the calculation of plasma equilibria in toroidal devices. The methodology involves the extension of quasilinear theory (QLT) to a periodic geometry. In view of Prof. Kaufman's very significant work on QLT [2], especially his formulation of QLT in axisymmetric toroidal configurations [3], the calculation is quite relevant to this Symposium.

The usual MHD equilibrium condition[22] $\boldsymbol{j} \times \boldsymbol{B} = \boldsymbol{\nabla} P$ predicts that $\boldsymbol{B} \cdot \boldsymbol{\nabla} P = 0$. For a system with good magnetic flux surfaces, this merely states that the pressure gradient is perpendicular to the flux surface. However, for regions with stochastic magnetic fields (of interest for tokamaks with ergodic limiters and for the edge regions of high-β stellarators), the condition is conventionally interpreted to mean that the pressure profile is flattened. Nevertheless, nonvanishing pressure gradients are observed in both experiments and simulations [75], so a more refined description is required.

Reiman *et al.* [75] observed that flattening need not occur in the presence of weak flows or an anisotropic pressure tensor. Furthermore, they outlined a hybrid analytical/numerical procedure wherein the magnetic-field configuration could be calculated without explicit reference to those effects. One begins with the quasineutrality condition $\boldsymbol{\nabla} \cdot \boldsymbol{j} = 0$. That can be rewritten as

$$\boldsymbol{B} \cdot \boldsymbol{\nabla} \mu = -\boldsymbol{\nabla} \cdot \boldsymbol{j}_\perp, \tag{65}$$

where $\mu \doteq j_\parallel / B$; this is a magnetic differential equation for μ ($\boldsymbol{j}_\perp$ is supposed to be given). Given μ, one can then proceed to solve Ampere's law for $\boldsymbol{B}$, for example by using the `PIES` code [76].

Let the magnetic field be written as $\boldsymbol{B} = \boldsymbol{B}_0 + \delta \boldsymbol{B}$, where $\boldsymbol{B}_0$ possesses good flux surfaces and $\delta \boldsymbol{B}$ is a stochastic perturbation. Equation (65) then takes the form

$$\boldsymbol{B}_0 \cdot \boldsymbol{\nabla} \mu + \delta \boldsymbol{B} \cdot \boldsymbol{\nabla} \mu = -\boldsymbol{\nabla} \cdot \boldsymbol{j}_\perp. \tag{66}$$

In coordinates for which the unperturbed magnetic lines are straight, one has

$$\boldsymbol{B}_0 \cdot \boldsymbol{\nabla} = B_0^\phi \left(\frac{\partial}{\partial \phi} + \iota \frac{\partial}{\partial \theta} \right), \tag{67}$$

where θ and ϕ are the (generalized) poloidal and toroidal angles and ι is the rotational transform. With the convention $\mu \sim \exp(\mathrm{i}m\theta - \mathrm{i}n\phi)$, Fourier series analysis of Eq. (67) leads to

$$(B_0^\phi)^{-1} \boldsymbol{B}_0 \cdot \boldsymbol{\nabla} \mu \to -\mathrm{i}(n - \iota m) \widehat{\mu}_{mn}, \tag{68}$$

demonstrating resonances at each of the rational surfaces. Those resonances are broadened by the stochastic component of the field. To treat the broadening quantitatively, one may think of $\delta \boldsymbol{B}$ as a random variable; then Eq. (66) becomes a multiplicatively nonlinear stochastic differential equation to which the usual statistical techniques may be applied. The problem is difficult because $\delta \boldsymbol{B}$ is self-consistently determined in terms of μ through Ampere's law. As a

[21] So recent, in fact, that the research was not performed until well after the KaufmanFest.

[22] There seems to be no area of plasma physics that Prof. Kaufman has not touched. One of his earliest papers dealt with stability of a "hydromagnetic" equilibrium [74].

first step, Krommes and Reiman [66] (KR) ignored that complication and treated Eq. (66) as a problem of *passive advection*. The resulting statistical closure contains the usual magnetic diffusion coefficient D_m (KR worked in the quasilinear limit), but also involves additional terms in principle because the source term $-\nabla \cdot \boldsymbol{j}_\perp$ is statistically correlated with $\delta \boldsymbol{B}$. This point can be illustrated with the aid of a model, as thoroughly discussed in Appendix A of Ref. 66. In the end, the additional terms can be argued to be small, so one is left with a resonance-broadened magnetic differential equation of the form

$$\boldsymbol{B}_0 \cdot \nabla \langle \mu \rangle - D_m \nabla^2 \langle \mu \rangle = -\nabla \cdot \langle \boldsymbol{j}_\perp \rangle. \tag{69}$$

(The notation "$D_m \nabla^2$" denotes a nontrivial operator, since the magnetic coordinates appropriate to the underlying flux surfaces are not orthogonal.)

This is not the end of the story. Equation (69) must be solved subject to the periodicity constraints in a torus; that requires an interesting foray into the theory of periodic Green's functions (reviewed in Appendix B of Ref. 66). Furthermore, Ampere's law requires the *total* $\mu = \langle \mu \rangle + \delta \mu$, where $\delta \mu$ is the fluctuation about the mean. Of course, D_m incorporates mean-square information about $\delta \mu$, but that is not sufficient for the solution of Ampere's law in a single stochastic realization. To address this issue, it is convenient to formulate the statistical closure in terms of a nonlinear Langevin equation. I will not describe the details here, since they are thoroughly discussed in Ref. 66 (see also Ref. 28). I merely remark that the nonlinear Langevin approach is undoubtedly the most intuitive formulation of the statistical closure problem, so it is highly recommended for the connoisseurs.

It was already recognized in Ref. 73 that the problem of stochastic magnetic fields presents an important venue for the application of statistical methods. The present calculation shows that the esoterica of renormalization can lead to quite practical, even quantitative predictions. Although this is well known in quantum electrodynamics, examples in plasma physics are few. One is therefore encouraged to pursue related avenues that connect analytical theory and experimental or simulation data. Some of those are described in the next several sections.

3. Tools for zonal-flow physics

The renormalized theory of spectral balance is quite general, but that is a weakness as well as a strength since it treats all fluctuations on a formally equal footing. For example, in homogeneous turbulence for which plane waves are sensible eigenfunctions, the theory describes coupling between all wave numbers $\boldsymbol{k} = (k_x, k_y, k_z)$, with no combination of Cartesian components being singled out for special treatment. Physics considerations, however, may point to a preferred role for specific $\boldsymbol{k}$'s.

Consider drift waves, for example, which are very important in fusion-related microturbulence. The classic drift wave has the dispersion relation

$$\Omega_{\boldsymbol{k}} = \frac{\omega_*(\boldsymbol{k})}{1 + k_\perp^2 \rho_s^2}. \tag{70}$$

Here the diamagnetic frequency is $\omega_* \doteq k_y V_*$, the diamagnetic velocity is $V_* \doteq (\rho_s / L_n) c_s$, the sound speed is $c_s \doteq (ZT_e/m_i)^{1/2}$ (Z is the atomic number), the sound radius is $\rho_s \doteq c_s/\omega_{ci}$, and the density scale length is $L_n \doteq (-\partial_x \ln\langle n \rangle)^{-1}$. This wave is obtained from the assumptions of fluid ions ($\omega / k_\parallel v_{ti} \gg 1$) and adiabatic electrons ($\omega / k_\parallel v_{te} \ll 1$). The $k_\perp^2 \rho_s^2$ term describes the shielding effect of ion polarization drift. If that is ignored, then the physics of the wave[23] is that $\boldsymbol{E} \times \boldsymbol{B}$ advection rearranges the background ion density profile, creating a fluctuation in ion

[23] For further discussion of drift- and zonal-flow physics, see a set of pedagogical lectures by Krommes [8].

charge density. That fluctuation is then neutralized essentially instantaneously by electron flow along the magnetic field lines (adiabatic electron response).

Clearly adiabatic response fails when $\omega/k_\parallel v_{te} \geq 1$. Such is the case for *convective cells* with $k_\parallel = 0$. In toroidal geometry, for which magnetic field lines do not in general close on themselves but instead ergodically fill a magnetic flux surface, the condition $k_\parallel = 0$ requires both $k_y = 0$ and $k_z = 0$. Thus wave numbers with $\boldsymbol{k} = (k_x, 0, 0)$ have a special role. Note that electrostatic potentials dependent on only the x coordinate give rise to $\boldsymbol{E} \times \boldsymbol{B}$ flows dominantly in the y direction: $\boldsymbol{V}_E \propto \widehat{\boldsymbol{b}} \times \boldsymbol{\nabla}\varphi \approx \widehat{\boldsymbol{z}} \times \boldsymbol{\nabla}\varphi = \widehat{\boldsymbol{y}}\,\varphi'(x)$. Such flows are called *zonal flows*. In general, they are sheared: $\varphi''(x) \neq 0$.

It has come to be appreciated that the physics of zonal flows is very important in the theory of tokamak microturbulence. (For a review, see Ref. 77.) From now on, I shall use $\boldsymbol{q} = (q_x, 0, 0)$ for the zonal-flow wave number. Zonal flows do not directly tap the free energy in the density gradient, since their diamagnetic frequency $\omega_*(\boldsymbol{q}) \propto q_y$ vanishes. They do not respond adiabatically, but rather exhibit fluid response. They can be driven up by nonlinear drift-wave–drift-wave interactions and can react back on the drift waves, both interactions being described by special cases of the triad relation $\boldsymbol{k} + \boldsymbol{p} + \boldsymbol{q} = \boldsymbol{0}$, where $\boldsymbol{k}$ and $\boldsymbol{p}$ denote drift waves. For example,

$$\underbrace{(k_x, k_y, k_z)}_{\boldsymbol{k}} + \underbrace{(k_x - q_x, -k_y, -k_z)}_{\boldsymbol{p}} + \underbrace{(q_x, 0, 0)}_{\boldsymbol{q}} = (0, 0, 0). \tag{71}$$

Because in general zonal flows are sheared and tend to be of long wavelength, they can shear apart drift-wave eddies, which tends to suppress the fluctuations that are directly responsible for the turbulent transport. Thus the presence of zonal flows can lead to reduced transport.

I shall discuss two calculations relating to zonal-flow physics, selected because they nicely illustrate a variety of fundamental tools. First I will remark on the problem of the transition to ion-temperature-gradient-driven (ITG) turbulence, in which the excitation of zonal flows plays a key role. That transition can be addressed by bifurcation theory, an important tool in the modern theory of nonlinear dynamics. Then I will mention facets of the calculation of the zonal-flow nonlinear growth rate, which affords an opportunity to apply multiple-scale methods, Hamiltonian methods for PDEs, and some interesting wave kinetics.

3.1. Bifurcation theory and the transition to ion-temperature-gradient-driven turbulence

I have remarked that zonal flows can shear apart drift-wave eddies, thereby suppressing them. Perhaps the most dramatic example of this effect is the so-called *Dimits shift*, which arises in the transition to ITG turbulence. In order to appreciate the Dimits shift, consider the most naive paradigm for the onset of turbulent transport. Imagine changing an order or bifurcation parameter ϵ in order to bring a system from linear stability, through marginal stability at $\epsilon = \epsilon_c$, to linear instability. (ϵ_c can be chosen to vanish.) Naive arguments might suggest that fluctuations, hence turbulent transport, should turn on as soon as ϵ exceeds ϵ_c. In the ITG problem, ϵ can be taken to be a normalized temperature gradient, $\epsilon \doteq R/L_T$ (R is the major radius of the tokamak). The conventional scenario is then sketched in Fig. 3.

However, large-scale numerical simulations due to Dimits and collaborators [78] showed that when collisions (which are weak) are neglected, the actual situation is more complicated. In fact, the transport does not turn on until ϵ exceeds some $\epsilon_* > \epsilon_c$. The difference $\Delta \doteq \epsilon_* - \epsilon_c$ is known as the *Dimits shift*. By qualitative analysis of the simulation results, the Dimits shift was understood in Ref. 78 to be due to the excitation of zonal flows, which then totally suppressed the ITG modes for $\epsilon_c < \epsilon < \epsilon_*$. Further discussion and insights were given by Rogers *et al.* [79].

However, more can be done. In the systematic theory of nonlinear dynamics, transitions at critical points such as $\epsilon = \epsilon_c$ are conventionally analyzed by means of bifurcation theory [80, 81], so it is natural to ask [82] whether the Dimits shift can also be explained in those terms. Now if

Journal of Physics: Conference Series **169** (2009) 012005 doi:10.1088/1742-6596/169/1/012005

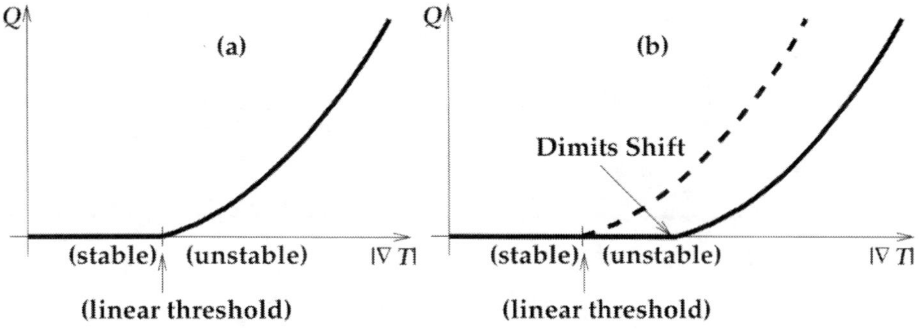

Fig. 3. Illustration of the Dimits shift. (a) Naive expectation: heat flux Q should turn on immediately after the threshold for linear instability is crossed. (b) Simulation result: turbulence and transport are suppressed in a nonzero interval (the "Dimits shift") above threshold.

zonal flows were not involved, the transition at $\epsilon = \epsilon_c$ would be a conventional Hopf bifurcation, in which a complex-conjugate pair of eigenvalues crosses the imaginary λ axis. (Perturbations are assumed to vary as $e^{\lambda t}$.) If the bifurcation is supercritical, nonzero drift-wave activity (not necessarily turbulent) would be expected for ϵ just above ϵ_c, but that is not observed. One must rule out subcritical bifurcation, but in fact it is clear (as I will explain) that the standard theory of Hopf bifurcation does not apply when zonal flows are involved.

To gain insight into this issue, Kolesnikov and Krommes (KK) analyzed a "simple" model of ITG fluctuations. For the details, which are lengthy and nontrivial, I refer the reader to Refs. 83 and 84. Here I will just describe the qualitative situation, which I hope will argue in favor of the use of fundamental and systematic analysis tools.

The basic model consists of two nonlinear 2D PDEs that couple the vorticity[24] ϖ and the pressure P. The nonlinearity is due to $\boldsymbol{E} \times \boldsymbol{B}$ advection. The system supports a linear ITG instability for sufficiently large background temperature gradient. The zonal modes are assumed to be undamped in order to make contact with the collisionless simulations reported by Dimits *et al.*

KK worked in a finite-sized box, the radial width of which was intended to model the width of a radial eigenfunction in the presence of magnetic shear. They imposed standing-wave boundary conditions in the radial direction; that is the only place in which magnetic shear entered the calculation. Periodic boundary conditions were used in the poloidal direction.

By Galerkin projection, KK reduced the problem to a system of coupled ODEs. They used the minimum number of modes that leads to a well-posed problem. One satisfies the triangle condition with modes $(k_x, k_y) = (1, 1)$, $(1, -1)$, and $(2, 0)$. The latter (zonal) amplifudes are real; the others are complex. Upon recalling that one has $n = 2$ coupled fields, one can count the number of real degrees of freedom by $2n + 2n + n = 10$. Thus one deals with nonlinear dynamics in a 10D phase space.

In bifurcation calculations, progress is made by analyzing the dynamics on the *center manifold*. That is the nonlinear generalization of the center eigenspace, namely the linear subspace spanned by the eigenvectors with zero real parts at marginal stability. In a standard Hopf bifurcation, the center eigenspace would be 2D. If there are no unstable eigenvalues additional to the nearly marginal ones (in other words, the complement of the center eigenspace is a stable eigenspace), the center manifold theorem [80, 81] states that near the bifurcation

[24] Here the vorticity is the component of the curl of the $\boldsymbol{E} \times \boldsymbol{B}$ velocity in the direction of the magnetic field. In suitably dimensionless variables, it can be written as $\varpi = \nabla_\perp^2 \varphi$, where φ is the electrostatic potential. For some pedagogical discussion of the vorticity in strongly magnetized plasmas, see Ref. 85, Fig. 2.

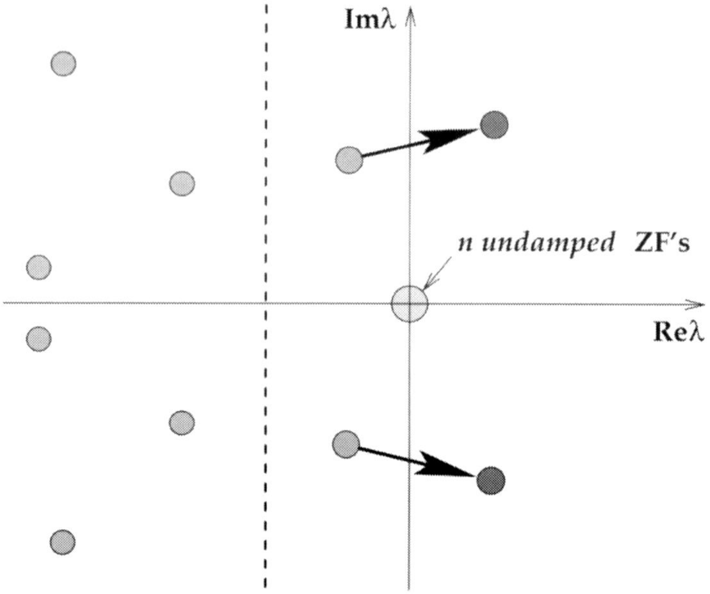

Fig. 4. (Color online.) Movement of eigenvalues as temperature gradient passes through the threshold for linear instability ($\partial_t \to \lambda$). Red: marginally stable drift waves; blue: strongly stable drift waves; yellow: undamped zonal flows.

point the nonlinear dynamics are attracted to a (curved) center manifold that is tangent to the center eigenspace at the bifurcation point. This behavior arises because the modes in the stable eigenspace are strongly damped, so rapidly relax and become adiabatically slaved to the center modes.

Thus, a standard Hopf bifurcation is associated with a 2D center manifold. The two degrees of freedom correspond to the amplitude ρ and phase θ in a complex representation $\rho e^{i\theta}$ of the reduced dynamics.

In the present model, the bifurcation is more complicated because the center eigenspace and the center manifold are 4D. This follows because each of the two (real) zonal modes that are retained contributes a linear eigenvalue of zero. Thus, the center eigenspace looks schematically like Fig. 4. If one ignores the drift-wave phase variable, the coordinate axes of the reduced phase space can be taken to be the two (real) zonal amplitudes and the drift-wave amplitude.

The ultimate goal is to understand the time-asymptotic dynamics on the center manifold. But to verify that one is on the right track, it is first useful to directly integrate the 10D system for a parameter value just slightly above marginality. A snapshot of the result is shown in Fig. 5, and a movie (`movie234.mpg`) is available with the online Proceedings. Those plot the dynamics versus two of the eight drift-wave degrees of freedom (on the x–y plane) and one of the zonal amplitudes (on the z axis). The phase point is projected at various time snapshots onto the horizontal and two back vertical planes; it is also drawn in three-space, with various temporal portions of the trajectory color-coded for easier visibility. It is seen that after an initial transient the trajectory settles down to a regular motion that subsequently spirals into a time-independent fixed point with zero drift-wave amplitudes but nonzero zonal amplitude, i.e., a pure zonal-flow state. The post-transient regular motion is evidence of a center manifold. The interpretation is consistent with our previous discussion: For arbitrary initial conditions that include excited drift-wave amplitudes, nonlinearity drives zonal flows that react back on the drift waves and ultimately suppress them entirely. This behavior occurs for a range of temperature gradients;

Journal of Physics: Conference Series **169** (2009) 012005 doi:10.1088/1742-6596/169/1/012005

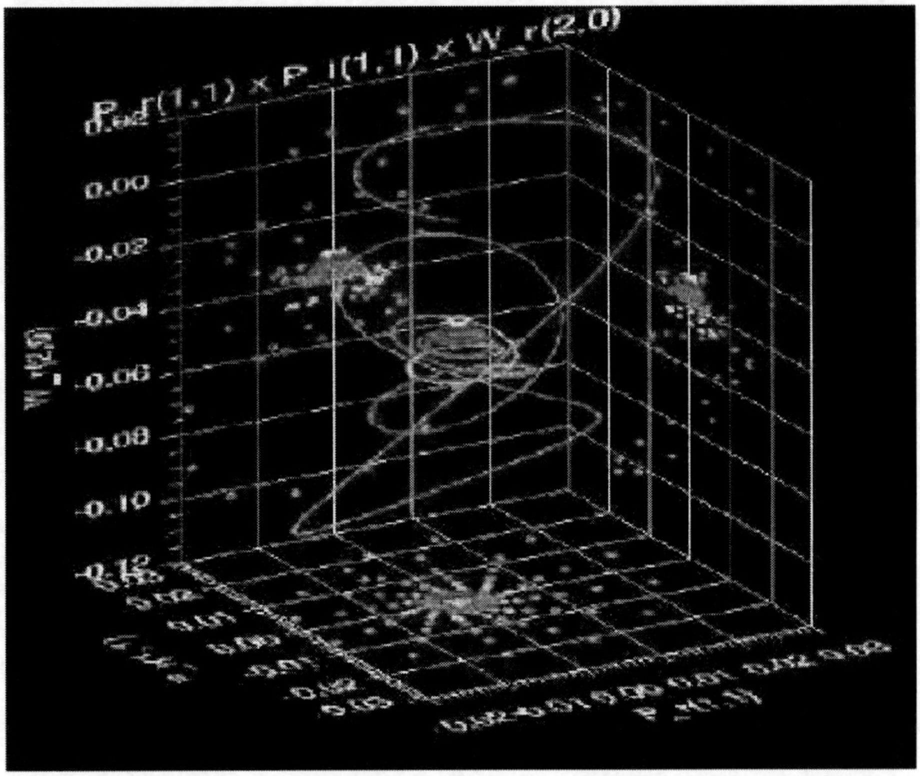

Fig. 5. (Color online.) Projection of the 10D trajectory into the three-space with axes $(x, y, z) = (\operatorname{Im} \varpi_{1,1}, \operatorname{Re} P_{1,1}, \varpi_{2,0})$. The time-asymptotic solution is a pure zonal-flow state.

that range defines the Dimits shift.

Analytically, one needs to (i) construct the center manifold; (ii) locate the fixed point; and (iii) discuss the stability of the fixed point and thus determine Δ, the width of the Dimits shift for the model. Construction of the center manifold is expedited by a projection method described by Kuznetsov [81], in which a contravariant coordinate system is used to locate points on the center manifold above the flat center eigenspace. (For a graph of this construction, see Ref. 84, Appendix B, Fig. 12. The projection method does not require the initial diagonalization of the linear matrix that was used in Ref. 80.) The construction is done perturbatively, thus is restricted to small amplitudes and valid only locally. Surprisingly, the location of the fixed point can be obtained analytically even though nonlinear algebraic equations need to be solved. Many details of the calculations, together with caveats[25] and lengthy discussion, may be found in Ref. 84.

The significance of this analysis is that it uses a systematic perturbation method to analyze a nontrivial and highly nonlinear dynamical problem. The reduction of a high-dimensional dynamics to the substantially lower-dimensional motion on the center manifold seems to be very much in the spirit of the calculations favored by Prof. Kaufman. Indeed, my initial introduction

[25] A significant problem of the present method is the severe Galerkin truncation. One could retain more modes, but then the center manifold becomes of higher dimensionality and difficult to analyze. Unfortunately, the predicted Dimits shift depends on the truncation. It may be that an entirely different functional analysis is required to properly calculate Δ. Certainly one desires an approach in which the influence of magnetic shear can be taken into account more precisely. There is plenty of room here for future work.

Journal of Physics: Conference Series **169** (2009) 012005 doi:10.1088/1742-6596/169/1/012005

to nonlinear dynamics came *via* the detailed study of Chirikov's thesis in the reading group in which I participated during my initial visits to Berkeley. (A subsequent additional important influence was the review article by Treve [34] on the possible applications of nonlinear dynamics ideas to controlled fusion research.) It is somewhat embarrassing that it took me almost 30 more years to accomplish a nontrivial calculation in nonlinear dynamics,[26] but better late than never.[27] Serious analytical theorists should certainly be taught the rudiments of that field.

3.2. Wave kinetics and the calculation of the zonal-flow growth rate

The bifurcation analysis emphasizes that zonal flows can suppress drift waves. However, although I have not demonstrated it here, the 10D model is not sophisticated enough to describe steady-state turbulence above the onset. There is considerable further work to do on "fundamental" studies of nonlinear dynamical behavior in fusion microturbulence. But one still has the general formalism of spectral balance in statistical closures. Here, I will discuss one aspect of that theory as it pertains to the generation of zonal flows in steady-state turbulence.

When physics is adequately described by a scalar field such as the potential φ, the general form of the spectral balance equation is Eq. (35), which I transcribe here for a Markovian, one-time description of convective cells with wave number $\boldsymbol{q}$:

$$\partial_t C_{\boldsymbol{q}} - 2\gamma_{\boldsymbol{q}}^{\mathrm{lin}} C_{\boldsymbol{q}} + 2\,\mathrm{Re}\,\eta_{\boldsymbol{q}}^{\mathrm{nl}} C_{\boldsymbol{q}} = 2F_{\boldsymbol{q}}, \tag{72}$$

which contains as usual a coherent nonlinear damping $\eta_{\boldsymbol{q}}^{\mathrm{nl}}$ and an incoherent nonlinear drive $F_{\boldsymbol{q}}$. In principle, both of these terms contain self-interactions, i.e., triadic couplings among three convective modes satisfying $\boldsymbol{q} + \boldsymbol{q}' + \boldsymbol{q}'' = \boldsymbol{0}$, as well as cross couplings between the convective cells and other constituents of the turbulence such as drift waves. Now for nonlinearities involving $\boldsymbol{E} \times \boldsymbol{B}$ advection, the mode-coupling coefficient includes a cross product: $M_{\boldsymbol{kpq}} \propto \widehat{\boldsymbol{z}} \cdot \boldsymbol{p} \times \boldsymbol{q}$. This implies that interactions between three zonal modes vanishes, since the wave vectors of all such modes point in the $\pm x$ direction and have vanishing cross product. There may be interactions between zonal modes and other kinds of convective cells. Here, however, I consider the important case where the zonal mode $\boldsymbol{q}$ interacts with two drift waves $\boldsymbol{k}$ and $\boldsymbol{p}$. Specifically, I discuss the calculation of the nonlinear zonal-flow growth rate $\gamma_{\boldsymbol{q}}^{\mathrm{nl}} \doteq -\,\mathrm{Re}\,\eta_{\boldsymbol{q}}^{\mathrm{nl}}$ due to that interaction.

Such a calculation was first attempted by Diamond and coworkers [89], who contributed important insights. Their method of attack was *via* a postulated wave kinetic equation for the drift waves as modulated by the presence of a zonal flow. That captures the right idea; however, some of the details were either incorrect or unclear.

A serious attempt to understand the wave kinetic approach was mounted by Krommes and Kim [69]; subsequently, further insights were gained by Krommes and Kolesnikov [90]. Those works together provide an interesting and nontrivial tour through a variety of useful tools, including renormalized statistical theory, Hamiltonian formulation of nonlinear PDEs, multiple-scale methods, and an unusual application of wave-kinetic analysis.

The first calculation performed by Krommes and Kim was a direct derivation of $\gamma_{\boldsymbol{q}}^{\mathrm{nl}}$ from the Markovian closure theory of the generalized Hasegawa–Mima equation[28] in the limit that $\epsilon \doteq q/k \ll 1$. In principle, this calculation is straightforward; one need "merely" write down the

[26] Early in my career I worked seriously on transport due to stochastic magnetic fields [73, 86], which required understanding of Chirikov's stochasticity criterion [71] and some issues relating to Kolmogorov entropy. However, the focus there was on statistical dynamics, not nonlinear dynamics in the usual sense.

[27] Examples of earlier work on bifurcations in plasmas include the papers by Hinton and Horton [87] and Horton *et al.* [88], although I do not believe that the concept of a center manifold was mentioned explicitly in those works.

[28] In the derivation of the original Hasegawa–Mima equation, adiabatic electron response was assumed. That is incorrect for convective cells, which exhibit fluid response. In the generalized Hasegawa–Mima equation, a modified Poisson equation is used in which the electrons do not respond at all for $k_{\parallel} = 0$ modes. For further discussion, see Ref. 69.

expression for $\eta^{\mathrm{nl}}_{\boldsymbol{q}}$, simplify the mode-coupling coefficients for small ϵ, and perform the indicated wave-number integrals. The rub is that those integrals, when constrained by the triad relation $\boldsymbol{q}+\boldsymbol{k}+\boldsymbol{p}=\boldsymbol{0}$, are over unusual, noncircular domains in $\boldsymbol{k}$ space. Specifically, if $\boldsymbol{q}$ is fixed, the tip of $\boldsymbol{k}$ and the tail of $\boldsymbol{p}$ are tied together, and that vertex is moved, then the vectors $\boldsymbol{k}$ and $\boldsymbol{p}$ sweep out areas that are not simultaneously symmetrical with respect to the same origin. Thus great care and cross checking is required, and the calculation took many months to complete.

One offshoot of this analysis was that the calculation of $\gamma^{\mathrm{nl}}_{\boldsymbol{q}}$ (done there in the plasma-physics context of the anisotropic generalized Hasegawa–Mima equation) was shown to be closely related to Kraichnan's earlier pioneering calculations of eddy viscosity for the isotropic 2D Navier–Stokes equation [33]. In fact, it was possible to formulate the plasma problem so that it embraced the Navier–Stokes equation as well, so one could take the appropriate limit. Bringing the various results into complete agreement provided an entirely nontrivial check on the calculations.

In the end, the formula for $\gamma^{\mathrm{nl}}_{\boldsymbol{q}}$ is fairly simple and closely related to the result of Diamond's heuristic wave-kinetic analysis (with some important details corrected and clarified). It is interesting to understand the underlying reasons why a wave-kinetic approach works in this context. Lest the reader argue that this is obvious, I point out that the calculations of $\gamma^{\mathrm{nl}}_{\boldsymbol{q}}$ have been performed for homogeneous turbulence, for which gradients of the statistical observables with respect to space vanish; the usual wave-kinetic analysis as applied to linear wave propagation involves the Poisson bracket $\{A, B\} = \partial_{\boldsymbol{X}}A\cdot\partial_{\boldsymbol{k}}B - \partial_{\boldsymbol{k}}A\cdot\partial_{\boldsymbol{X}}B$, which vanishes identically for a homogeneous background.

The resolution of this paradox is relatively easy. In calculating $\gamma_{\boldsymbol{q}}$ for a specific $\boldsymbol{q}$, one is essentially considering a conditional average. A selected $\boldsymbol{q}$ appears to the drift waves as a long-wavelength, inhomogeneous modulation. That modulation can be taken as the inhomogeneous background on top of which a wave-kinetic formalism can be built. This idea can be systematized by using the functional apparatus introduced by MSR, as demonstrated by Krommes and Kim.

Since the wave-kinetic formalism is being used in an unfamiliar context, one must rethink all of its assumptions very carefully. In particular, it is not immediately clear what "action" density should be evolved by the wave kinetic equation. In traditional wave theory, the appropriate action is (electrostatically) $\mathcal{J} \doteq (\partial\mathcal{D}/\partial\Omega)(E^2/8\pi)$ (Ω being the mode frequency). However, this is built from properties of the linear waves. It is unclear why linear wave physics should have anything directly to do with the calculation of $\gamma^{\mathrm{nl}}_{\boldsymbol{q}}$, which follows from properties of the *nonlinear* interaction, and indeed it is inappropriate to use $\mathcal{J}$. Instead, one must use an invariant of the drift-wave–zonal-flow interaction.

For generalized Hasegawa–Mima dynamics, the relevant invariant was first found by Smolyakov and Diamond [91], who essentially studied the inhomogeneous interaction entirely in $\boldsymbol{k}$ space. However, we know from the work of McDonald and Kaufman on the systematic analysis of weakly inhomogeneous systems with the aid of the Weyl calculus (for a review, see Ref. 7) that it is better to work in a mixed $\boldsymbol{x}$-space–$\boldsymbol{k}$-space representation. That naturally introduces Poisson brackets, the properties and interpretation of which are well known. Actually, the lowest-order version of such a calculation for weakly inhomogeneous statistical dynamics was first performed by Carnevale and Martin [68], although Krommes and Kim identified a fundamental conceptual mistake in that work. In any event, when all of the various approaches are brought into agreement, the Smolyakov–Diamond invariant is recovered. For the generalized Hasegawa–Mima equation, that is

$$\mathcal{N} = \sum_{\boldsymbol{k}} \mathcal{N}_{\boldsymbol{k}}, \quad \mathcal{N}_{\boldsymbol{k}} \doteq \frac{1}{2}(1+k^2)^2\,|\varphi_{\boldsymbol{k}}|^2\,. \tag{73}$$

It is important that this differs from the energy

$$\mathcal{E} = \frac{1}{2}\sum_{\boldsymbol{k}}(1+k^2)\,|\varphi_{\boldsymbol{k}}|^2\,. \tag{74}$$

At this point, Krommes and Kim proceeded to discuss the derivation of $\gamma_{\boldsymbol{q}}^{\rm nl}$ from a wave-kinetic equation that evolved $\mathcal{N}_{\boldsymbol{k}}$. However, the fundamental significance of $\mathcal{N}_{\boldsymbol{k}}$ remained unclear. Therefore, let me take a historical detour and jump ahead a few years. In the course of my work with Kolesnikov on the ITG problem, we were led to consider the generalization of the wave-kinetic approach to situations involving several coupled fields and possibly multiple nonlinear invariants. Those are complicated enough that it is essential to work with the most concise and systematic methodology available. Morrison's Hamiltonian formalism for nonlinear PDEs [92] is ideal for this purpose. For specific kinds of nonlinearity, happily including $\boldsymbol{E}\times\boldsymbol{B}$ advection, it allows one to derive the nonlinear portion of the PDE (or system of PDEs) from a particular Poisson bracket of the field with a Hamiltonian functional $\mathcal{H}$. Now although $\mathcal{H}$ is conserved when all nonlinear interactions are considered, it is not conserved for a subclass of fluctuations such as short-scale drift waves because energy can be transferred to other modes. However, in a bracket formalism it is known that additional *Casimir invariants* can be conserved for an arbitrary Hamiltonian due to particular properties of the Poisson bracket. Using a covariant bracket formalism combined with multiple-scale analysis, Krommes and Kolesnikov [90] were able to show that for $\boldsymbol{E}\times\boldsymbol{B}$ advection a particular Casimir is conserved even under the interaction between short-wavelength drift waves and long-wavelength convectives cells. For the generalized Hasegawa–Mima equation, the Smolyakov–Diamond invariant is recovered. For some more general systems of PDEs, multiple Casimirs may exist.

Multiple-scale analysis leads to a wave-kinetic description of the DW–CC interaction. In the face of multiple Casimirs, that formalism is interesting and nontrivial [90], but it is sufficient here to merely consider the case of a single invariant $\mathcal{N}$ (discussed at length in Ref. 69). Then the appropriate wave kinetic equation is

$$\partial_T\widetilde{\mathcal{N}}_{\boldsymbol{k}} - \{\widetilde{\Omega}_{\boldsymbol{k}}, \widetilde{\mathcal{N}}_{\boldsymbol{k}}\} = 0, \tag{75}$$

where the Poisson bracket is defined by Eq. (48) and $\widetilde{\Omega}_{\boldsymbol{k}}$ is an appropriate *nonlinear* advection frequency of the short scales by the long scales *(it is not the linear mode frequency of a short-scale drift wave)*. A tilde emphasizes that the quantity is random (due to the stochastic nature of the zonal flows). For the generalized Hasegawa–Mima equation, one has $\widetilde{\Omega}_{\boldsymbol{k}} = \boldsymbol{k}\cdot\widetilde{\boldsymbol{V}}_{\boldsymbol{q}}$.

When Eq. (75) is summed over $\boldsymbol{k}$ and integrated over $\boldsymbol{X}$, the Poisson bracket vanishes, confirming that $\mathcal{N}$ is truly invariant under the interaction between the long and the short scales. Energy $\widetilde{\mathcal{E}}$, however, is not invariant, since it is a different wave-number weighting of $\langle\delta\varphi^2\rangle_{\boldsymbol{k}}$ than is $\widetilde{\mathcal{N}}_{\boldsymbol{k}}$ and the Poisson bracket involves derivatives with respect to $\boldsymbol{k}$. One finds

$$\partial_T\widetilde{\mathcal{E}}_{\boldsymbol{k}} = \{\widetilde{\Omega}_{\boldsymbol{k}}, \widetilde{\mathcal{E}}_{\boldsymbol{k}}\} + 2\widetilde{\gamma}_{\boldsymbol{k}}^{(1)}\widetilde{\mathcal{E}}_{\boldsymbol{k}}, \tag{76}$$

where, if $\widetilde{\mathcal{N}}_{\boldsymbol{k}} = c_{\boldsymbol{k}}\widetilde{\mathcal{E}}_{\boldsymbol{k}}$,

$$\widetilde{\gamma}_{\boldsymbol{k}}^{(1)} \doteq \frac{1}{2}\{\widetilde{\Omega}_{\boldsymbol{k}}, c_{\boldsymbol{k}}\}. \tag{77}$$

It is important to realize that this first-order growth rate is *not* the desired long-wavelength growth rate in the statistical spectral balance equation; note that $\widetilde{\gamma}_{\boldsymbol{k}}^{(1)}$, being linearly proportional to the long-wavelength amplitude (buried in $\widetilde{\Omega}_{\boldsymbol{k}}$), vanishes when averaged over an ensemble of convective cells whereas the spectral balance equation holds after all averaging has been performed. To obtain $\gamma_{\boldsymbol{q}}^{\rm nl}$, one must proceed to second order. It is unnecessary to repeat the remainder of the calculation (which uses the wave kinetic equation a second time) here since the details have been described in a number of places [8, 69, 93]. (Ref. 8, in particular, makes a strong attempt to be introductory and pedagogical.) But the flavor of the calculation is clear, and one can now appreciate that systematic multiple-scale analysis and wave-kinetic formalism can be usefully applied not just to linear problems, but to nonlinear ones as well.

4. Current challenges in statistical turbulence theory: Intermittency and coherent structures

The well-established two-point moment closures do a reasonable job of describing both fluctuations in the energy-containing part of the wave-number spectrum and the associated turbulent transport.[29] However, they can fail badly at describing higher-order moments [100, 101]. In this section, I introduce some of the recent issues concerning highly non-Gaussian statistics. Few, if any, questions will be resolved here; this is a fertile area for future research.

4.1. Higher-order statistics

PDFs can (usually) be described by their infinity of moments [102] or, better, cumulants [21]. A Gaussian PDF $P(x)$ is described by just two cumulants: the mean $\overline{x} \doteq \langle x \rangle$, and the variance $\sigma^2 \doteq \langle \delta x^2 \rangle$. Basic measures of non-Gaussianity include the *skewness* S and *kurtosis* K (or the flatness factor F)[30]:

$$S \doteq \frac{\langle \delta x^3 \rangle}{\sigma^3}; \quad K \doteq F - 3, \text{ with } F \doteq \frac{\langle \delta x^4 \rangle}{\sigma^4}. \tag{78}$$

K is defined such that it vanishes for Gaussian statistics.

Kurtosis is a measure of the importance of large events. A PDF with large kurtosis frequently has an exponential or stretched exponential tail, i.e., more slowly decaying than a Gaussian. For example, the pure exponential PDF

$$P(x) = \frac{1}{2\sigma'} \mathrm{e}^{-|x|/\sigma'} \tag{79}$$

has $K = 3$. Realizations sampled from such PDFs display occasional and intermittent large spikes.

That kurtosis and intermittency are linked follows from a simple example given by Frisch [103, Chap. 8]. Consider a time series $\widetilde{x}(t)$ constructed by turning on a random signal $x(t)$ only a fraction λ of the time. The kurtosis of $\widetilde{x}$ is

$$K = \frac{\lambda \langle x^4 \rangle}{(\lambda \langle x^2 \rangle)^2} - 3 = \lambda^{-1}(\langle x^4 \rangle / \langle x^2 \rangle^2) - 3, \tag{80}$$

which approaches infinity as $\lambda \to 0$. [For a Gaussian signal, the formula reduces to $3(1-\lambda)/\lambda$.]

One cannot construct PDFs by specifying S, K, and other cumulants arbitrarily. Cumulants must satisfy an infinity of *realizability constraints*, which stem from the positive-semidefiniteness of a valid PDF. The simplest one is $\langle \delta x^2 \rangle \geq 0$; another is

$$K \geq S^2 - 2. \tag{81}$$

This constraint will be important in the next section.

Various recent plasma experiments clearly display intermittent statistics. Time series taken from probes inserted in the edges of various tokamaks are routinely intermittent. Visual evidence can be found from gas-puff-imaging (GPI) data [104], which shows the intermittent formation of density *blobs* that are occasionally ejected into the scrape-off layer (SOL) and may propagate outward to the wall. A snapshot is shown in Fig. 6. Although a good deal of understanding of the

[29] The justification for this assertion is the extensive work done in the 1990's by Bowman [94–96] and Hu [97–99] on realizable Markovian statistical closures. That work included successful comparisons between the predictions of statistical closure theory and direct numerical simulations.

[30] It is unfortunate that the names of the kurtosis and flatness have not been entirely standardized in the literature. My definition of F (I follow Frisch [103, p. 41]) is often confusingly called the kurtosis, and my definition of K is sometimes called the "excess kurtosis."

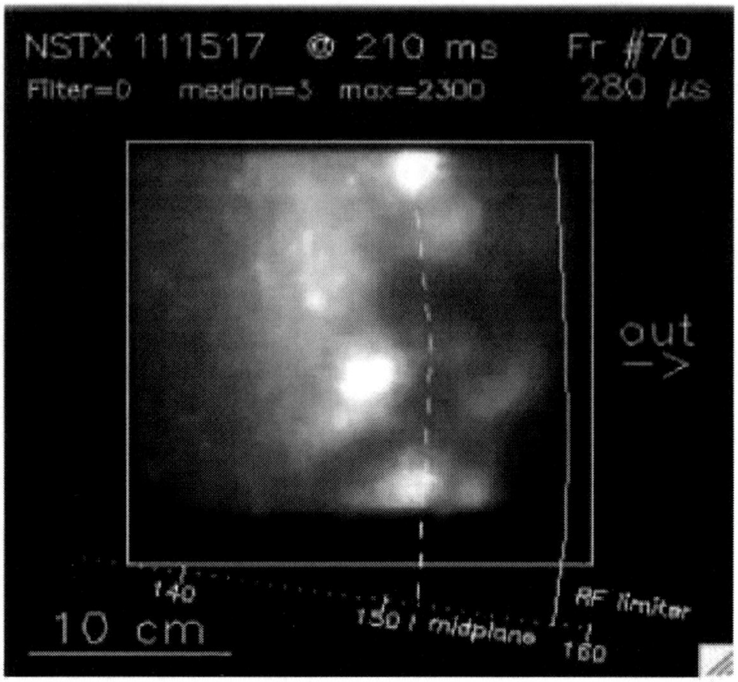

Fig. 6. (Color online.) Fluctuations in the edge of NSTX show the intermittent formation of "blobs." From `http://www.pppl.gov/~szweben/NSTX04`.

propagation of a single blob has been developed,[31] blob generation is much more complicated because blobs arise intermittently from a sea of turbulence. Research on blob generation is ongoing [106]; it requires a comprehension of many facets of plasma dynamics as well as statistical turbulence theory.

4.2. Physical relationships between kurtosis and skewness

A more basic physics experiment may shed light on intermittency in plasmas. In the machine TORPEX (`http://crppwww.epfl.ch/torpex/`), skewness and kurtosis have been measured at many places in the machine in a large ensemble of shots [107]. The data shows the clear relationship

$$K = aS^2 + b, \tag{82}$$

with $a \approx 1.5$ and $b \approx -0.22$. These values are consistent with the realizability constraint (81). However, no physics basis for this relationship was offered by the original investigators.

Strikingly, essentially the same relationship, *with the same value of* a, is evinced by a global database of fluctuations in sea-surface temperature, as shown by Sura and Sardeshmukh (SS) [108]. Furthermore, those investigators advanced an analytical calculation, based on a nonlinear Langevin equation, that predicted the values $a \approx \frac{3}{2}$ and $b \approx 0$. This is clearly suggestive.

The essence of the SS model is the inclusion of a white-noise multiplicative nonlinearity in the evolution equation for the fluctuations:

$$\partial_t \delta T + \cdots + \sigma \widetilde{f}(t) \delta T = \widetilde{f}(t), \tag{83}$$

[31] The basic physics of blob propagation in a torus is $\boldsymbol{E} \times \boldsymbol{B}$ motion due to internal polarization arising from magnetic drifts [105].

where $\widetilde{f}$ is Gaussian white noise and σ measures the strength of the nonlinearity. Given such a stochastic differential equation [109] (to be analyzed in the Stratonovich interpretation), a Fokker–Planck equation can be deduced. The structure of that equation is simple enough that a closed hierarchy of analytical relations can be obtained for the moments through a given order. Eliminating the variance in the equations for K and S gives rise to the relation (82).

The validity of the model even for its intended target of sea-surface temperature fluctuations may be questionable; however, its prediction of $a = \frac{3}{2}$ is obviously a nontrivial success. It is reasonable, therefore, to inquire whether a similar approach may be fruitful for the application to TORPEX. Such a calculation was first outlined in Ref. 110, and some details have been published [111]. Here I shall just mention a few of the salient issues.

4.2.1. Multiplicative white noise and Markovian Langevin equations. The most important step in the modeling is the representation of all nonlinear effects by the product $\widetilde{f}\,\delta T$, where $\widetilde{f}$ is the external noise that models the excitation of the turbulence (in lieu of a term involving the linear growth rate). Assuming that $\widetilde{f}$ is white ensures a Markovian (time-local) description. Similar descriptions are well known in statistical closure theory in the form of *realizable Langevin representations* for closures such as the EDQNM. In those theories, the effects of the original quadratic nonlinearity are represented by a coherent nonlinear damping $\eta^{\rm nl}\delta\psi$ and an *internal* incoherent drive $\widetilde{f}^{\rm nl} \propto \widetilde{w}(t)$, where $\widetilde{w}$ is Gaussian white noise. However, that $\widetilde{w}$ is specifically *not the same* as an external stirring force $\widetilde{f}$, so one must dig deeper to understand the connection, if any, to the SS model. The remainder of the discussion becomes too technical for inclusion here; I refer the interested reader to Ref. 111.

4.2.2. The role of linear waves. The SS model applies to a single real scalar field. In confined plasmas, however, linear waves are ubiquitous. Even a simple drift mode, linearly described by $\partial_t\varphi + \mathrm{i}\Omega\varphi = 0$, must be represented by a 2D system of first-order ODEs that couple the real and imaginary parts of the complex wave amplitude. Not only does this increase the complexity of the analysis, the presence of a real frequency has consequences for the ultimate statistics of the associated nonlinear equation. If one follows the spirit of the SS model, it can be shown that although the basic scaling $K = aS^2 + b$ still holds, one has, in particular, $a = a(\Omega)$. For the derivation of that relation for a simple model, see Refs. 110 and 111.

4.3. What can one learn from single-point statistics?

The simplest model that includes the effects of wave propagation predicts that the variation of a with Ω is bounded and modest; the principal effect of nonzero Ω is to reduce S and K. That may account for some of the variability in the data. Presumably shots with large values of skewness are highly nonlinear and relatively insensitive to a linear wave frequency; such shots should tend to track the $a = \frac{3}{2}$ curve, as indeed they appear to.

Ultimately, what does one learn from single-point statistics? At present, the answer is unclear. The level of the current modeling, which is not derived from realistic PDEs for the plasma physics, is too crude to enable one to deduce fine points of the underlying dynamics, no matter how suggestive is the $\Omega = 0$ prediction $a \approx \frac{3}{2}$. It remains possible that a more refined theory of the PDF for density fluctuations obeying specific nonlinear PDEs can be developed. The gold standard here is the *mapping closure* technique due to Chen *et al.* [50]. Unfortunately, that theory is very difficult (see additional discussion and references in Ref. 28), particularly when applied to coupled PDEs that contain linear waves. Extensive further development is required.

In addition to the theoretical complexity of systematic PDF theories, it must be stressed that knowledge of the single-point PDF of a fundamental scalar field such as temperature is not sufficient for predictions of transport coefficients, which require cross correlations. Those

Journal of Physics: Conference Series **169** (2009) 012005 doi:10.1088/1742-6596/169/1/012005

are easily available from moment closures, which however have a great deal of difficulty in describing intermittency. It may be that entirely new tools are required in order to make progress. Certainly numerical investigations will continue to play an increasingly important role in explorations of complicated nonlinear statistical dynamics, but additional analytical methods would be welcome as well. As younger generations develop those, they could do no better than to look for background and motivation to the elegant tools, and the physical insights behind them, to which the present Symposium pays tribute.

5. Summary and discussion

In the previous sections, I have tried to provide a snapshot of some of the interesting issues in modern plasma turbulence theory, emphasizing connections to, or generalizations of, some of the tools that were either developed by Prof. Kaufman and/or his students and colleagues or, in my opinion, would appeal to their instincts for systematology and elegance. Although turbulent plasmas in the laboratory are incredibly complex, the nonlinear equations that describe them are overly rich with dynamical possibilities, and many questions remain unanswered, it does appear that a predictive superstructure is beginning to appear.

The short title of this article is "Tools for fusion-related microturbulence." Although I mentioned several fusion-related applications (toroidal plasma equilibria in stochastic magnetic fields, and the Dimits shift), fusion-specific details have been discussed here only minimally. How then can the title be justified, especially when calculations aimed specifically at practical fusion phenomena such as the Dimits shift can be done only for extremely idealized models? My answer is that the more complicated the practical details, the more the field benefits from pedagogical exemplars in the sense of Kuhn [29]. (For a short discussion of Kuhn's paradigm concept, see Ref. 112.) Hopefully the methods and models discussed above will help the reader structure his or her understanding of the basic logical threads underlying the behavior of turbulent plasma confinement devices and will point the way to future analytic calculations that are yet more concise and elegant.

In Regis's book *Who Got Einstein's Office?* [113], he comments on the vision of Abraham Flexner, the self-styled educator and author of the influential 1910 Flexner Report on the state of American medical education: "[Flexner] dreamt of a hybrid university—or an institute. . . —in which faculty members and students went into virgin territories together, not as equals, perhaps, but at least as partners." Of course, the physical realization of that dream is Princeton's Institute for Advanced Study. But the description of Flexner's ideal interaction between faculty members and students strikes me as perfectly applicable to the intellectual environment that Allan Kaufman fostered at Berkeley. He and his students, post-docs, and visitors together tackled many difficult and subtle problems in theoretical physics, distilling new and elegant principles and perspectives that have created recognizable order from the chaos of myriad physical effects and made it so much easier for the newcomer to comprehend the basic structure of plasma dynamics. I feel privileged to have had even a brief taste of plasma-physics research *a la* Kaufman. That Allan's influence has not faded is evidenced by the fact that in organizing my thoughts for this Symposium I was led to new results on renormalized ponderomotive force, 30 years after the original version of the K–χ theorem was published. I suggest that an aspiring theorist could do no better than to study the body of Allan's work; it will inspire one and lead to the truth.

Finally, I would like to note the influence that my early Berkeley experiences had on my own teaching philosophy. I never sat in Allan's classroom, for which I am very sorry. However, his approach to physics is very compatible with my own inclination, which is to seek out overarching principles that speak to the fundamental unity of the myriad of nonlinear phenomena. One must certainly teach specific known facts, but only by embedding those into a more general logical framework can one progress to generalizations and new discoveries. In my own teaching, I try to emphasize that framework, including insights that stem from, or were motivated by, discoveries made by Allan and his collaborators. I am sure that I do it much less well than Allan, but I am very grateful indeed for his inspiration. I am deeply honored to have been able to participate in *Plasma Theory, Wave Kinetics, and Nonlinear Dynamics: KaufmanFest 2007.*

Acknowledgments

This work was supported by U.S. Department of Energy Contract No. DE–AC02–76–CHO–3073. Some parts of this material were initially presented orally at (i) a December 8, 2004, Colloquium at Princeton University's Plasma Physics Laboratory, and (ii) the June 11–12, 2007, Symposium on Recent Advances in Plasma Physics held at Princeton University in honor of Prof. R. C. Davidson. I am grateful to R. Dewar for discussions of renormalized Lie perturbation theory; to S. Johnston for providing me with some historical background on the K–χ theorem; to A. Reiman for collaboration on the problem of equilibria in stochastic magnetic fields; and to M. Labit, P. Sura, T. Stoltzfus-Dueck, and S. Zweben for discussions of blob physics and higher-order statistics.

References

[1] Kevorkian J K and Cole J D 1996 *Multiple Scale and Singular Perturbation Methods (Applied Mathematical Sciences, Vol. 114)* (New York: Springer)

[2] Kaufman A N 1972 *J. Plasma Phys.* **8** 1–5

[3] Kaufman A N 1972 *Phys. Fluids* **15** 1063–9

[4] Dewar R L 1973 *Phys. Fluids* **16** 1102–7

[5] Dewar R L 1976 *J. Phys. A* **9** 2043–57

[6] Cary J R and Kaufman A N 1977 *Phys. Rev. Lett.* **39** 402–4

[7] McDonald S W 1988 *Phys. Rep.* **158** 337–416

[8] Krommes J A 2006 *Turbulence and Coherent Structures in Fluids, Plasmas and Nonlinear Media* ed Shats M and Punzmann H (Singapore: World Scientific) pp 115–232

[9] Vedenov A, Velikhov E and Sagdeev R 1962 *Proceedings of the Conference on Plasma Physics and Controlled Nuclear Fusion Research (Salzburg, 1961) [Nucl. Fusion Suppl. Pt. 2]* (Vienna: International Atomic Energy Agency) pp 465–75 translated in U.S.A.E.C. Division of Technical Information document AEC–tr–5589 (1963), pp. 204–37.

[10] Drummond W E and Pines D 1962 *Proceedings of the Conference on Plasma Physics and Controlled Nuclear Fusion Research (Salzburg, 1961) [Nucl. Fusion Suppl. Pt. 3]* (Vienna: International Atomic Energy Agency) pp 1049–57

[11] Kadomtsev B B 1965 *Plasma Turbulence* (New York: Academic Press) translated by L. C. Ronson from the 1964 Russian edition *Problems in Plasma Theory*, edited by M. A. Leontovich. Translation edited by M. C. Rusbridge

[12] Sagdeev R Z and Galeev A A 1969 *Nonlinear Plasma Theory* (New York: W. A. Benjamin)

[13] Benney D J and Newell A C 1969 *Stud. Appl. Math.* **48** 29–53

[14] Davidson R C 1972 *Methods in Nonlinear Plasma Theory* (New York: Academic Press)

[15] Dupree T H 1966 *Phys. Fluids* **9** 1773–82

[16] Dupree T H 1967 *Phys. Fluids* **10** 1049–55

[17] Kraichnan R H 1959 *J. Fluid Mech.* **5** 497–543

[18] Kraichnan R H 1961 *J. Math. Phys.* **2** 124–48 erratum: J. Math. Phys. **3**, 205 (1962)

[19] Orszag S A and Kraichnan R H 1967 *Phys. Fluids* **10** 1720–36

[20] Wyld H W 1961 *Ann. Phys.* **14** 143–65

[21] Kubo R 1962 *J. Phys. Soc. Japan* **17** 1100–20

[22] Lorenz E N 1963 *J. Atmos. Sci.* **20** 130–41

[23] Kraichnan R H 1967 *Phys. Fluids* **10** 1417–23

[24] Chirikov B V 1969 Research concerning the theory of non-linear resonance and stochasticity Nuclear Physics Institute of the Siberian Section of the USSR Academy of Sciences Report 267, translated by A. T. Sanders (CERN Trans. 71–40)

[25] Balescu R 1960 *Phys. Fluids* **3** 52–63

[26] Lenard A 1960 *Ann. Phys.* **10** 390–400

[27] Weinstock J 1968 *Phys. Fluids* **11** 1977–81

[28] Krommes J A 2002 *Phys. Rep.* **360** 1–351

[29] Kuhn T 1970 *The Structure of Scientific Revolutions* 2nd ed (Chicago: U. of Chicago Press) with postscript

[30] Candlestickmaker S 1972 *Q. Jl R. astr. Soc.* **13** 63–6

[31] Martin P C, Siggia E D and Rose H A 1973 *Phys. Rev. A* **8** 423–37

[32] Orszag S A 1977 *Fluid Dynamics* ed Balian R and Peube J L (New York: Gordon and Breach) pp 235–374

[33] Kraichnan R H 1976 *J. Atmos. Sci.* **33** 1521–36

[34] Treve Y M 1978 *Topics in Nonlinear Dynamics: A Tribute to Sir Edward Bullard* ed Jorna S (New York: AIP) p 147

[35] Rose H A 1979 *J. Stat. Phys.* **20** 415–47

[36] Benford G and Thomson J J 1972 *Phys. Fluids* **15** 1496–502

[37] Smith G R and Kaufman A N 1975 *Phys. Rev. Lett.* **34** 1613–6

[38] Krommes J A and Oberman C 1976 *J. Plasma Phys.* **16** 229–60

[39] Johnston S and Kaufman A N 1978 *Phys. Rev. Lett.* **40** 1266–9

[40] DuBois D F and Espedal M 1978 *Plasma Phys.* **20** 1209–39

[41] Krommes J A 1978 *Theoretical and Computational Plasma Physics* (Vienna: International Atomic Energy Agency) pp 405–17

[42] Littlejohn R G 1979 *J. Math. Phys.* **20** 2445–58

[43] Dubin D H E, Krommes J A, Oberman C R and Lee W W 1983 *Phys. Fluids* **26** 3524–35

[44] Brizard A J and Hahm T S 2007 *Rev. Mod. Phys.* **79** 421–68

[45] Krommes J A 2007 *Phys. Plasmas* **14** 090501/1–26

[46] Ford J 1083 *Phys. Today* **36** 40–7

[47] Ford J 1983 *Long-time Prediction in Dynamics* ed Horton Jr C W, Reichl L E and Szebehely V G (New York: Wiley) p 79

[48] Holmes P 1990 *Phys. Rep.* **193** 137–63

[49] Diacu F and Holmes P 1996 *Celestial encounters: The origins of chaos and stability* (Princeton, NJ: Princeton University Press)

[50] Chen H D, Chen S and Kraichnan R H 1989 *Phys. Rev. Lett.* **63** 2657–60

[51] Kraichnan R H 1991 *New Perspectives in Turbulence* ed Sirovich L (New York: Springer) chap 1, p 1

[52] Holmes P, Lumley J L and Berkooz G 1996 *Turbulence, Coherent Structures, Dynamical Systems and Symmetry* (Cambridge: Cambridge University Press)

[53] Kraichnan R H 1962 *Proc. Symp. Appl. Math.* **13** 199

[54] Bender C M and Orszag S A 1978 *Advanced Mathematical Methods for Scientists and Engineers* (New York: McGraw–Hill)

[55] Zinn-Justin J 1996 *Quantum Field Theory and Critical Phenomena* 3rd ed (Oxford: Oxford University Press)

[56] Barenblatt G I 1996 *Scaling, Self-similarity, and Intermediate Asymptotics* (Cambridge: Cambridge University Press)

[57] Schwinger J 1951 *Proc. Natl. Acad. Sci.* **37** 452–5

[58] Schwinger J 1951 *Proc. Natl. Acad. Sci.* **37** 455–9

[59] Martin P C 1976 private communication

[60] Kraichnan R H and Montgomery D 1980 *Rep. Prog. Phys.* **43** 547–619

[61] Kaufman A N 1984 *Phys. Lett. A* **100** 419–22

[62] Morrison P 1984 *Phys. Lett. A* **100** 423–7

[63] Deprit A 1969 *Cel. Mech.* **1** 12–30

[64] Kaufman A N 1987 *Phys. Rev. A* **36** 982–4

[65] Krommes J A and Kotschenreuther M T 1982 *J. Plasma Phys.* **27** 83–94

[66] Krommes J A and Reiman A H 2008 *Phys. Plasmas* In preparation

[67] Orszag S A 1969 *Turbulence of Fluids and Plasmas* ed Fox J (Brooklyn, New York: Polytechnic Press) p 17

[68] Carnevale G F and Martin P C 1982 *Geophys. Astrophys. Fluid Dynamics* **20** 131–64

[69] Krommes J A and Kim C B 2000 *Phys. Rev. E* **62** 8508–39

[70] Cary J R 1981 *Physics Rep.* **79** 129–59

[71] Lichtenberg A J and Lieberman M A 1992 *Regular and Chaotic Dynamics* 2nd ed (New York: Springer)

[72] Krommes J A 1984 *Handbook of Plasma Physics* vol 2 ed Galeev A A and Sudan R N (Amsterdam: North–Holland) chap 5.5, pp 183–268

[73] Krommes J A, Oberman C and Kleva R G 1983 *J. Plasma Phys.* **30** 11–56

[74] Newcomb W A and Kaufman A N 1961 *Phys. Fluids* **4** 314–24

[75] Reiman A, Zarnstorff M C, Monticello D, Weller A, Geiger J and the W7–AS Team 2007 *Nucl. Fusion* **47** 572–8

[76] Reiman A H and Greenside H 1986 *Comput. Phys. Commun.* **43** 157–67

[77] Diamond P H, Itoh S I, Itoh K and Hahm T S 2005 *Plasma Phys. Control. Fusion* **47** R35–R161

[78] Dimits A M, Bateman G, Beer M A, Cohen B I, Dorland W, Hammett G W, Kim C, Kinsey J E, Kotschenreuther M, Kritz A H, Lao L L, Mandrekas J, Nevins W M, Parker S E, Redd A J, Shumaker D E, Sydora R and Weiland J 2000 *Phys. Plasmas* **7** 969–83

[79] Rogers B N, Dorland W and Kotschenreuther M 2000 *Phys. Rev. Lett.* **85** 5336–9

[80] Guckenheimer J and Holmes P 1983 *Nonlinear Oscillations, Dynamical Systems, and Bifurcations of Vector Fields* (New York: Springer)

[81] Kuznetsov Y A 1998 *Elements of Applied Bifurcation Theory* (New York: Springer)

[82] Kolesnikov R A 2005 *Bifurcation theory of the transition to collisionless ion-temperature-gradient-driven plasma turbulence* Ph.D. thesis Princeton U.

[83] Kolesnikov R A and Krommes J A 2005 *Phys. Rev. Lett.* **94** 235002/1–4

[84] Kolesnikov R A and Krommes J A 2005 *Phys. Plasmas* **12** 122302/1–25

[85] Krommes J A 2006 *Mathematical and Physical Theory of Turbulence: Proceedings of the International Turbulence Workshop* ed Cannon J and Shivamoggi B (Boca Raton, FL: Chapman & Hall/CRC) pp 69–90

[86] Krommes J A 1978 *Prog. Theor. Phys. Suppl.* **64** 137–47

[87] Hinton F L and Horton Jr C W 1971 *Phys. Fluids* **14** 116–23

[88] Horton W, Hu G and Laval G 1996 *Phys. Plasmas* **3** 2912–23

[89] Diamond P H, Rosenbluth M N, Hinton F L, Malkov M, Fleischer J and Smolyakov A 1998 *17th IAEA Fusion Energy Conference* (Vienna: International Atomic Energy Agency) pp 1421–8 IAEA–CN–69/TH3/1

[90] Krommes J A and Kolesnikov R A 2004 *Phys. Plasmas* **11** L29–L32

[91] Smolyakov A I and Diamond P H 1999 *Phys. Plasmas* **6** 4410–3

[92] Morrison P J 1998 *Rev. Mod. Phys.* **70** 467–521

[93] Krommes J A 2004 *J. Plasma Fusion Res. SERIES* **6** 84–8

[94] Bowman J C 1992 *Realizable Markovian Statistical Closures: General Theory and Application to Drift-wave Turbulence* Ph.D. thesis Princeton University

[95] Bowman J C, Krommes J A and Ottaviani M 1993 *Phys. Fluids B* **5** 3558–89

[96] Bowman J C and Krommes J A 1997 *Phys. Plasmas* **4** 3895–909

[97] Hu G 1995 *Statistical Theory of Resistive Drift-wave Turbulence and Transport* Ph.D. thesis Princeton University

[98] Hu G, Krommes J A and Bowman J C 1995 *Phys. Lett. A* **202** 117–25

[99] Hu G, Krommes J A and Bowman J C 1997 *Phys. Plasmas* **4** 2116–33

[100] Chen H, Herring J R, Kerr R M and Kraichnan R H 1989 *Phys. Fluids A* **1** 1844–54

[101] Krommes J A 1996 *Phys. Rev. E* **53** 4865–82

[102] Carleman T 1922 *C. R. Acad. Sci.* **174** 1680–2

[103] Frisch U 1995 *Turbulence* (Cambridge: Cambridge University Press)

[104] Maqueda R J, Wurden G A, Stotler D P, Zweben S J, LaBombard B, Terry J L, Lowrance J L, Mastrocola V J, Renda G F, D'Ippolito D A, Myra J R and Nishino N 2003 *Rev. Sci. Instrum.* **74** 2020–6

[105] Krasheninnikov S I 2001 *Phys. Lett. A* **283** 368–70

[106] Stoltzfus-Dueck T, Krommes J A and Zweben S J 2007 *Bull. Am. Phys. Soc.* **52(16)** 303

[107] Labit B, Furno I, Fasoli A, Diallo A, Müller S H, Plyushchev G, Podestà M and Poli F M 2007 *Phys. Rev. Lett.* **98** 255002/1–4

[108] Sura P and Sardeshmukh P D 2007 *J. Phys. Oceanogr.* **38** 638–46

[109] Gardiner C W 2004 *Handbook of Stochastic Methods* 3rd ed (Berlin: Springer)

[110] Krommes J A 2007 *Bull. Am. Phys. Soc.* **52(16)** 370

[111] Krommes J A 2008 *Phys. Plasmas* **15** 030703/1–4

[112] Bird A 2007 (fall) *The Stanford Encyclopedia of Philosophy* ed Zalta E N (U. of Stanford) `http://plato.stanford.edu/archives/fall2007/entries/thomas-kuhn/` Sec. 3, The Paradigm Concept

[113] Regis E 1987 *Who got Einstein's office?* (Reading, MA: Addison–Wesley)

Journal of Physics: Conference Series **169** (2009) 012006 doi:10.1088/1742-6596/169/1/012006

Thoughts on brackets and dissipation: old and new

P J Morrison

Department of Physics and Institute for Fusion Studies, The University of Texas at Austin, Austin, TX 78712

E-mail: `morrison@physics.utexas.edu`

Abstract.
Bracket formulations of two kinds of dynamical systems, called incomplete and complete, are reviewed and developed, including double bracket and metriplectic dynamics. Dissipation based on the Cartan-Killing metric is introduced. Various examples of incomplete and complete dynamics are discussed, including dynamics associated with three-dimensional Lie algebras.

1. Introduction

The idea of extending the Hamiltonian or Lagrangian framework to include dissipation dates to Rayleigh, but active interest in this idea was renewed in 1981. At that time, Allan Kaufman had the idea to formulate dissipation in terms of brackets with some algebraic properties, an extension of the work on noncanonical Poisson brackets that I had begun a couple of years earlier with John Greene [1, 2]. Allan invited me to join him, and this resulted in a paper on quasilinear theory [3], which unfortunately is the only paper I have published with Allan. By 1984 we had gone our separate ways on this topic ([4, 5, 6, 7, 8]). Others became interested too (e.g. then [9] and afterwards [10, 11]) and, in particular, in two papers [12, 13] the double bracket for describing a particular kind of dissipation was introduced. In this work I will review some of the above, and describe and develop a few old ideas I had back in the mid 1980's, but never published.

The point of a bracket formulation is to imbue a dynamical system with certain structural properties that are generally of a geometrical flavor. For example, the ordinary Poisson bracket ensures the invariance of a closed, nondegenerate two-form, and the geometrical ramifications, such as the existence of the Poincare' invariants, that follow from this. Dynamical systems theory is a huge endeavor, encompassing mappings, a variety of ordinary and partial differential equations, and other more exotic equations. Here, encompassing statements applicable to finite and infinite-dimensional systems are sought, and thus the discussion need be of a general and formal nature. Two kinds of systems are described: incomplete and complete, a terminology similar to but different from the open, closed, and isolated systems of thermodynamics. Incomplete systems are ones that are not complete, but of special interest here are systems described by gradient dynamics of some type, while complete systems are described by metriplectic dynamics, a terminology introduced in [7]. Because incomplete and complete systems may have a Hamiltonian component, in the remainder of this introduction some features of Hamiltonian systems are described first, as needed for the discussions of incomplete and complete systems that follow.

Journal of Physics: Conference Series **169** (2009) 012006 doi:10.1088/1742-6596/169/1/012006

Hamiltonian systems serve as the archetype of energy conserving nondissipative systems, and can be viewed as a dynamical extension of the first law of thermodynamics. Here the Hamiltonian will be a time-independent function, $H(z)$, that can be identified with energy. The Hamiltonian dynamics considered will be noncanonically Hamiltonian, which for finite-dimensional systems means it has the form

$$\dot{z}^i = [z^i, H] = J^{ij}\frac{\partial H}{\partial z^j}, \qquad i,j = 1,2,\ldots,m, \tag{1}$$

where repeated up and down indices are summed, $z = (z^1, z^2, \ldots, z^m)$ denotes phase space coordinates, and the Poisson bracket is defined by

$$[f,g] = \frac{\partial f}{\partial z^i} J^{ij} \frac{\partial g}{\partial z^j}, \qquad i,j = 1,2,\ldots m, \tag{2}$$

for some cosymplectic form, $J(z)$, is a Lie algebra realization over $\mathbb{R}$, i.e. it is (i) bilinear, (ii) antisymmetric, and (iii) satisfies the Jacobi identity. Much has been written about this kind of dynamics and the interested reader is referred to [14, 15, 16]. Canonical Hamiltonian systems are the special case where $J = J_c$ with

$$J_c = \begin{pmatrix} 0_n & I_n \\ -I_n & 0_n \end{pmatrix}, \tag{3}$$

where $n = m/2$ denotes the number of degrees of freedom, 0_n denotes an $n \times n$ block of zeros, and I_n is the $n \times n$ identity.

Associated with noncanonical Hamiltonian systems are two kinds of invariants: those that commute with the particular H of interest, which will be denoted by $\mathcal{P} = \{P_1(z), P_2(z), \ldots\}$, where each P_i satisfies $[\mathcal{P}, H] = 0$, and a special class of invariants, distinct from $\mathcal{P}$, that are built into the phase space. These invariants, called Casimir invariants and denoted by $\mathcal{C} = \{C_1(z), C_2(z), \ldots\}$, satisfy $[\mathcal{C}, f] \equiv 0$ for any phase space function f. Thus they are invariants for any Hamiltonian. Casimir invariants require $\det J \equiv 0$, while $H \in \mathcal{P}$ is always the case.

For canonical Hamiltonian systems, a phase space point z_e is an equilibrium point iff it satisfies $\partial H/\partial z^i|_{z_e} = 0$ for all $i = 1, 2, \ldots n$. For noncanonical Hamiltonian systems, the Hamiltonian is not unique because $F := H + \sum_i \lambda_i C_i$ produces the same equations of motion for any $\lambda_i \in \mathbb{R}$ when inserted in the Poisson bracket. However, different equilibria are obtained from $\partial F/\partial z^i|_{z_e} = 0$ for all $i = 1, 2, \ldots n$, for different choices of the λ_i.

Equations that describe macroscopic media in terms of Eulerian variables are noncanonically Hamiltonian systems. This includes Euler's equations for fluid motion, magnetohydrodynamics, the Vlasov equation, etc. (see e.g. [15, 17] and many references therein).

The *Incomplete systems* of interest here dissipate energy and possibly other physical quantities. Energy, $E(z)$, leaves the system at a rate prescribed at each phase space point z. Thus, $\dot{E}(z) \leq 0$ for all z. Systems governed by gradient flows (see e.g. [18]), which for finite m-dimensional systems have the form

$$\dot{z} = -\nabla_z E, \tag{4}$$

where $\nabla_z = (\partial/\partial z^1, \partial/\partial z^2, \ldots, \partial/\partial z^m)$, are incomplete systems. Evidently,

$$\dot{E} = -|\nabla_z E|^2 \leq 0, \tag{5}$$

and so by Lyapunov's theorem, gradient flows have built-in asymptotic stability, i.e. a temporal relaxation to an equilibrium point z_e that satisfies $\partial E/\partial z^i|_{z_e} = 0$. This follows because E serves as a Lyapunov function.

Journal of Physics: Conference Series **169** (2009) 012006 doi:10.1088/1742-6596/169/1/012006

Stokes flow, which describes motion in the low Reynolds number limit of the the Navier-Stokes equation, is a physical example of an infinite-dimensional gradient flow. Other examples of infinite-dimensional gradient flows are Cahn-Hilliard systems [19] and Ricci flows [20, 21], which are nonlinear diffusion-like equations. By Serrin's theorem [22], the asymptotic stability of Stokes flow also occurs for the full nonlinear Navier-Stokes equation, provided the Reynolds number is small enough, yet this system, because it has a Hamiltonian component, is not a gradient flow. This suggests that there may be some more general structure lurking. Indeed, this is the subject matter of Sec. 2, where generalizations of gradient flows are described.

As basic theories, incomplete systems are incomplete: they do not properly account for the lost energy, which would require coupling to the outside dynamics. The system's loss is the outside's gain, and the gain is not accounted for in a dynamical sense, but just assumed to vanish in a prescribed way. They are also incomplete because they do not describe the dynamics of both energy and entropy. One or the other does not enter into the formulation in a physically precise manner. Nevertheless, such systems that do not conserve energy or ignore entropy (or vice versa) are quite useful in engineering and practical applications, the Navier-Stokes equation being a case in point. Also, dissipative dynamics can be useful for proving theorems and constructing numerical algorithms.

Complete systems conserve energy and produce entropy. They can be viewed as dynamical extensions of thermodynamics that possess built-in structure emblematic of the first and second laws of thermodynamics and, as such, possess both Hamiltonian and dissipative components. The first law of energy conservation is embodied in the requirement that H be conserved under the full dynamics. It is also assumed that the set $\mathcal{P}$ is conserved. In addition, complete systems are assumed to possess an entropy function, $S(z)$, that satisfies $\dot{S}(z) \geq 0$ for all z, which is an embodiment of the second law. Thus the system may produce entropy, but it is possible there could be motions where entropy remains constant.

As defined, complete systems do not exchange entropy, energy, or other physical quantities with the outside. Indeed one could say there is no outside. For partial differential equations this would mean, depending on the topology of the domain, the existence of certain boundary conditions at infinity. However, to avoid being overly restrictive for infinite degree-of-freedom systems (e.g. partial differential equations) this is relaxed to allow for finite boundaries by only requiring local conservation of all physical quantities except those that are entropy-like. With this alteration of the definition, boundary fluxes of $\mathcal{P}$ may exist and entropy could also traverse the boundary. Archetype complete systems are the Boltzmann equation and the Vlasov equation with collisions [4, 5, 7], which conserve mass, momentum and energy, but produce entropy.

Complete systems with no boundary fluxes are kindred dynamical extensions of the isolated systems of thermodynamics, systems that do not exchange heat, work, or matter with an environment. The closed systems of thermodynamics exchange energy but not matter with an environment. Complete systems can accomplish this by suitable boundary conditions on $\mathcal{P}$ and the allowance of boundary entropy fluxes. Similarly, complete systems can embody the open systems of thermodynamics that allow, in addition to heat and work, matter fluxes. Incomplete systems do not naturally line up with ideas from thermodynamics because either entropy or energy dynamics is missing in the formulation.

In Sec. 3 metriplectic dynamics, the paradigmatic complete system, is discussed. As stated, metriplectic systems possess a dynamical first law of conservation of energy and a second law of entropy production. It will be seen that the Casimir invariants of the noncanonical Hamiltonian description are candidate entropy functions and the Hamiltonian plays the role of the thermodynamic internal energy. A shortcoming of metriplectic formulation is the apparent stitching together of the Hamiltonian and dissipative parts. An attempt to build a more unified structure is considered in Sec. 4, where a special kind of dissipation based on the Cartan-Killing metric of a Lie algebra is introduced. This metric can be utilized in the context of incomplete or complete systems. Some special cases corresponding to the three-dimensional Lie algebras are worked out. Finally, in Sec. 5, we conclude and summarize.

2. Incomplete Systems

To begin with, consider the general finite-dimensional incomplete system of the form

$$\dot{z}^i = [z^i, H] + \mathcal{D}^i(z)\,, \qquad i = 1, 2, \ldots, m\,, \tag{6}$$

where $\mathcal{D}$ represents dissipation of some kind. Analogous infinite-dimensional systems are written as

$$u_t = [u, H] + \mathcal{D}(u)\,, \tag{7}$$

where $u(x, t)$ depends on independent variables $x \in \Omega$ and $t \in \mathbb{R}$, and

$$[F, G] = \int_\Omega dx \frac{\delta F}{\delta u} \mathcal{J} \frac{\delta G}{\delta u}\,, \tag{8}$$

where $\mathcal{J}$ is a cosymplectic operator and $\delta F/\delta u$ denotes the function derivative of the arbitrary functional F with respect to the dependent variable u (see e.g. [14, 15]). It is of interest to construct 'collision operators' i.e. dissipative terms $\mathcal{D}$ that relax to steady states for future times, $t \geq 0$. In Sec. 2.1 one way to do this is shown [23], while in Sec. 2.2 the double bracket formulation of [12, 13] is described.

2.1. *Constructing H-theorems: general projection*

Consider a partial differential equation of the form (7). It is not difficult to construct dissipative terms $\mathcal{D}$ that formally relax to solutions associated with the Hamiltonian system without the dissipation. The Boltzmann H-theorem serves as a guide. Because the Hamiltonian part conserves $\mathcal{P} = \{P_1(z), P_2(z), \ldots\}$, the quantity

$$E := \sum_i \lambda_i P_i\,, \tag{9}$$

where the $\lambda_i \in \mathbb{R}$ are arbitrary, satisfies the following when $\mathcal{D} \neq 0$:

$$\frac{dE}{dt} = \int_\Omega dx \frac{\delta E}{\delta u} u_t = \int_\Omega dx \frac{\delta E}{\delta u} \mathcal{D}\,. \tag{10}$$

This suggests an obvious choice for $\mathcal{D}$, viz. $\mathcal{D}_1 := -\delta E/\delta u$, which means (7) has the form of a gradient flow and

$$\frac{dE}{dt} = -\int_\Omega dx \left(\frac{\delta E}{\delta u}\right)^2 \leq 0\,. \tag{11}$$

Whence one infers that the system could relax to states that satisfy $\delta E/\delta u = 0$.

As an example consider the KdV equation with dissipation,

$$u_t + u u_x + u_{xxx} = \mathcal{D}_1\,, \tag{12}$$

and recall Kruskal's soliton variational principle $\delta E/\delta u = 0$ with

$$E = \lambda_0 P_0 + \lambda_1 P_1 + \lambda_2 P_2 \tag{13}$$

where

$$P_0 = \int_{\mathbb{R}} dx\, u\,, \quad P_1 = \int_{\mathbb{R}} dx \frac{u^2}{2}, \quad P_2 = \int_{\mathbb{R}} dx \left(\frac{u^3}{6} - \frac{u_x^2}{2}\right)\,. \tag{14}$$

Extrema of this version of his principle are the single soliton solutions. Assuming

$$\mathcal{D}_1 = -\eta\frac{\delta E}{\delta u}\,, \tag{15}$$

where η is a parameter that sets the time scale for the relaxation, gives the following equation of motion:

$$u_t + uu_x + u_{xxx} = -\eta\left(\lambda_0 + \lambda_1 u + \lambda_2\frac{1}{2}u^2 + \lambda_2 u_{xx}\right)\,. \tag{16}$$

Here the first three terms on the right correspond to quadratically non-linear damping and the last is Burger's dissipation. When evaluated on the single soliton solution the right-hand-side vanishes, as of course does the left.

Generalizing a bit, suppose $\mathcal{D}_2 = -\mathcal{L}\delta E/\delta u$, where $\mathcal{L}$ is a non-negative operator, i.e. it satisfies $\langle\phi, \mathcal{L}\phi\rangle = ||T\phi||^2 \geq 0$ for all ϕ for some pairing or inner product. It is easy to construct such operators by simply writing $\mathcal{L} = \mathcal{A}^\dagger\mathcal{A}$ which could be a positive operator, i.e. satisfy $\langle\phi, \mathcal{L}\phi\rangle > 0$ for all $\phi \neq 0$, or it could have some desired degeneracies built-in that result in the conservation of some of the P_i's in the presence of this dissipation. With this choice

$$\frac{dE}{dt} = -\int_\Omega dx\,\left(\mathcal{A}\frac{\delta E}{\delta u}\right)^2 \leq 0\,. \tag{17}$$

Similarly, as a final possibility consider $\mathcal{D}_3 = -\mathcal{A}^\dagger\mathcal{K}\mathcal{A}\delta E/\delta u$, where $\mathcal{K}$ is some other known positive or non-negative operator.

The above constructions suggest a degenerate metric form for dissipation. For finite systems this would be

$$\dot{z}^i = [z^i, H] + (z^i, E)\,, \qquad i = 1, 2, \ldots, m\,, \tag{18}$$

with

$$(f, g) = -\frac{\partial f}{\partial z^i}g^{ij}\frac{\partial g}{\partial z^j}\,, \qquad i, j = 1, 2, \ldots m\,, \tag{19}$$

where the 'metric' $g_{ij}(z)$ is symmetric and non-negative, i.e. it has positive eigenvalues and is possibly degenerate ($\det g = 0$).

There is freedom in two respects with the above construction: one can choose E to obtain desired solutions to which the system relaxes, and one can design the degeneracies of g to preserve certain elements of $\mathcal{P}$. For example, g can be constructed using a projection operator so that for some $P \in \mathcal{P}$, $g^{ij}\partial P/\partial z^j = 0$ for all i. Then if $E = H$, the equation of motion follows from

$$\dot{z}^i = [z^i, E] + (z^i, E)\,, \qquad i = 1, 2, \ldots m\,, \tag{20}$$

and the system can relax to solutions of $\partial(H + P)/\partial z^i|_{z_e} = 0$. Evidently, the gradient flows of Sec. 1 correspond to the case where g is the Kronecker delta, i.e. $g^{ij} = \delta_{ij}$.

2.2. Double bracket dynamics

If the operator $\mathcal{A}$ of Sec. 2.1 above is chosen to be the cosymplectic form, $\mathcal{J}$, then a Casimir preserving relaxation to equilibria z_e that satisfy $\partial F/\partial z^i|_{z_e} = 0$ is obtained. This nice idea was introduced in [12, 13], and this special form of dissipative bracket is now referred to as the double bracket, which for finite-dimensional systems has the form

$$\{\{f, g\}\} = \sum_{k=1}^{m} J^{ik}J^{jk}\frac{\partial f}{\partial z^i}\frac{\partial g}{\partial z^j}\,. \tag{21}$$

Journal of Physics: Conference Series **169** (2009) 012006 doi:10.1088/1742-6596/169/1/012006

The equations of motion follow from

$$\dot{z}^i = [z^i, F] + \{\{z^i, F\}\}, \qquad i = 1, 2, \ldots m, \tag{22}$$

where $[\,,\,]$ is a noncanonical Poisson bracket with Casimir invariants $\mathcal{C}$ and $F = H + \mathcal{C}$ (a notation that means F is the sum of H plus a linear combination of elements of $\mathcal{C}$). Evidently,

$$\dot{\mathcal{C}} = \{\{\mathcal{C}, F\}\} = 0, \tag{23}$$

for any $\mathcal{C}$, and

$$\dot{F} = \{\{F, F\}\} = \{\{H, H\}\} \leq 0. \tag{24}$$

As an example, consider again the KdV equation which is noncanonically Hamiltonian with the Gardner bracket [24],

$$[F, G] = -\int_\Omega dx \frac{\delta F}{\delta u}\frac{\partial}{\partial x}\frac{\delta G}{\delta u}, \tag{25}$$

and the Hamiltonian P_2 of (14). The only Casimir of the Gardner bracket is P_0 of (14). The double bracket based on (25) is

$$\{\{F, G\}\} = \int_\Omega dx \frac{\delta F}{\delta u}\frac{\partial^2}{\partial x^2}\frac{\delta G}{\delta u} \tag{26}$$

and the resulting equation of motion is

$$u_t + uu_x + u_{xxx} = \left(\frac{\delta H}{\delta u}\right)_{xx} = (uu_x + u_{xxx})_{xx} \tag{27}$$

When $\Omega = \mathbb{R}$ and the boundary conditions $u = 0$ at $\pm$ infinity is assumed, this results in the trivial solution.

However, in the case of the ideal fluid, the situation is much richer and this double bracket construction has been used as a means of numerically obtaining equilibrium solutions that preserve the Casimir invariants for Euler's fluid equations [13] and for calculating the V-states of contour dynamics [17]. It is noted in passing that the double bracket formalism has been explored in a deeper and quite interesting sense in the context of the Toda lattice [25].

3. Complete Systems

Metriplectic dynamics [7], a systemization and formalization of the work of Allan Kaufman, the author, and others, describes complete systems. As stated in Sec. 1, it is a dynamic generalization of thermodynamics that embodies both the first law of conservation of energy and a second law of internal entropy production. Because of this, it could be argued that it is a paradigm for the most basic classical dynamics. The structure of metriplectic dynamics is similar to that above for incomplete systems, in that metriplectic dynamics possess both Hamiltonian and dissipative components, but the dynamics is generated by both a Hamiltonian and an entropy function and the metric is designed to have degeneracies of a particular type.

The metriplectic equation of motion has the form

$$\dot{z}^i = \{z^i, F\}_M = [z^i, F] + (z^i, F)_M \tag{28}$$

where $[\,,\,]$ is a noncanonical Poisson bracket and $F = H + S$, where S is the entropy function. As stated in Sec. 1, Casimir invariants are candidate entropies, i.e. S is chosen from $\mathcal{C}$, and this choice determines the equilibrium state z_e. Given that Casimir invariants are related to relabeling symmetries [26, 27], and thus a counting of states, it is not too surprising that entropies should come from $\mathcal{C}$. As before, $[f, \mathcal{C}] = 0$ implies

$$[f, S] = 0 \qquad \text{for all f}, \tag{29}$$

and, as in (19), dissipation is described by the bracket

$$(f, g)_M = \frac{\partial f}{\partial z^i} g_M^{ij} \frac{\partial g}{\partial z^j}, \qquad i, j = 1, 2, \dots m, \tag{30}$$

but with a sign change and the following additional requirement:

$$(f, \mathcal{P})_M = 0 \qquad \text{for all f}. \tag{31}$$

Thus upon identifying H with the energy and the entropy S with an element of $\mathcal{C}$, the dynamical embodiment of the first and second laws are as follows:

$$\dot{H} = \{H, F\}_M = [H, F] = 0 \qquad \text{and} \qquad \dot{S} = \{S, F\}_M = (S, S)_M \geq 0. \tag{32}$$

As noted above, the brackets for this have been worked out for many systems, including kinetic theories [4, 5, 7, 8], the compressible Navier-Stokes equation [6], and other systems [9, 10, 11]. Here the case of the free rigid body as given in [7] is briefly reviewed.

The free rigid body noncanonical Poisson bracket has Lie-Poisson form (e.g. [15, 16]), denoted $[f, g]_{LP}$, which means its cosymplectic form has the form

$$J^{ij} = c_k^{ij} z^k, \tag{33}$$

with c_k^{ij} being the structure constants of some Lie algebra. For classical theories this is generally a real Lie algebra. If $\{e^1, e^2, \dots, e^m\}$ denotes an m-dimensional Lie algebra basis, then the structure constants satisfy

$$[e^i, e^j]_{Lie} = c_k^{ij} e^k. \tag{34}$$

where $[\,,\,]_{Lie}$ is the Lie algebra nonassociative product. (Note, for consistency our index placement is dual to what is typical.) It can be shown that the Jacobi identity for the Poisson bracket defined by (33) follows from that for $[\,,\,]_{Lie}$.

For the rigid body [7], the Lie algebra is $so(3)$ with $c_k^{ij} = \epsilon_{ijk}$, the Levi-Civita symbol, and the coordinates $z = (z^1, z^2, z^3)$ correspond to the three components of the body (principal axes) frame angular momenta. The Hamiltonian is the rotational kinetic energy

$$H = \frac{1}{2} A_{ij} z^i z^j \qquad i, j = 1, 2, 3, \tag{35}$$

where $A = \text{dia}(1/I_1, 1/I_2, 1/I_3)$ is a diagonal matrix that depends on the principal moments of inertia, I_i. The Casimir invariant is proportional to the square of the angular momentum

$$C = \frac{1}{2} \delta_{ij} z^i z^j \qquad i, j = 1, 2, 3, \tag{36}$$

which serves as the entropy S, and the bracket $(f, g)_M$ is given by

$$(f, g)_M = - \sum_{i,j,k=1}^{3} \left[\frac{\partial H}{\partial z^i} \frac{\partial H}{\partial z^j} - \delta_{ij} \frac{\partial H}{\partial z^k} \frac{\partial H}{\partial z^k} \right] \frac{\partial f}{\partial z^i} \frac{\partial g}{\partial z^j}, \tag{37}$$

where we have scaled out η, the parameter that sets the time scale for relaxation. Observe (37) is a simple projection designed so that $\partial H / \partial z$ is a null eigenvector of g_M.

4. Cartan-Killing Dissipation

Metriplectic dynamics and its concomitant metriplectic manifold defined in [7] have an unnatural quality: despite the complementary alignment of the degeneracies of J and g, there is no deep intrinsic gemetrical connection between the Hamiltonian and dissipative components. One would hope for a structure with both components arising out of some common geometrical principle. As it stands, metriplectic dynamics has a clumsy and piecemeal quality with its brute force projection method, as evidenced by the lack of tensorial form of (37). For incomplete systems the double bracket formalism addresses this deficiency somewhat, in that the dissipation relies in an essential way on the cosymplectic form. However, (21) also lacks tensorial form and has an unnatural quality.

An origin of the problem is that symplectic manifolds have no natural metric, while for dissipation described by a gradient flow a metric is the essential ingredient. Obviously (31) would benefit from a natural metric. However, in addition, both the projection of (37) for complete systems and the double bracket expression of (21) for incomplete systems can be viewed as incorporating a Euclidean metric, and if a natural metric were available then they could be replaced by the following:

$$(f,g)_M = -\frac{\partial H}{\partial z^m}\frac{\partial H}{\partial z^n}\left[g^{mi}g^{nj} - g^{nm}g^{ij}\right]\frac{\partial f}{\partial z^i}\frac{\partial g}{\partial z^j}, \tag{38}$$

$$\{\{f,g\}\} = J^{in}g_{mn}J^{jn}\frac{\partial f}{\partial z^i}\frac{\partial g}{\partial z^j}. \tag{39}$$

Noncanonical Hamiltonian systems with Lie-Poisson brackets arise from a reduction based on symmetry (e.g. [15, 16]) and consequently are described by a particular Lie algebra. Thus, the idea comes to mind to incorporate the trace form used by Cartan to classify Lie algebras as a natural metric for dissipation. The trace form, sometimes called the Cartan-Killing metric, is defined in coordinates by

$$g^{ij}_{CK} = c^{in}_m c^{jm}_n, \tag{40}$$

where the structure constants $c^{ij}_k \in \mathbb{R}$ are defined as in Sec. 3. So, a dissipation naturally mated to a Lie-Poisson bracket appears to be

$$(f,g)_{CK} = \frac{\partial f}{\partial z^i} g^{ij}_{CK}\frac{\partial g}{\partial z^j}, \qquad i,j = 1,2,\ldots m. \tag{41}$$

This choice comes, so to speak, 'as is'. Because it is essentially a kinematic construction, there is no a priori guarantee that $(f,H)_{CK} = 0$ for all f. Thus a bracket of the form $\{f,g\} = [f,g]_{LP} + (f,g)_{CK}$ might not work, and for a metriplectic formulation one would need to use a bracket of the form of (38). Similarly, in general $(f,H)_{CK} \neq 0$ for all f, and so for an incomplete description of dissipation one might need to use (39). In any event, both these constructions seem to be superior.

Another problem with (41), a potential major problem, is that the eigenvalues of g_{CK} can have either sign. Because g_{CK} may also have zero eigenvalues, there could be expanding, contracting, and null directions, depending on the choices of H and S. It is possible that this seeming disadvantage could be turned into an advantage in selecting out preferential systems that possess the desired degeneracies, in which case, $\{f,g\} = [f,g]_{LP} + (f,g)_{CK}$ would work.

For semisimple Lie algebras, $\det g \neq 0$ and an inverse, g^{CK}_{jk}, can be defined, i.e.

$$g^{ik}_{CK}g^{CK}_{kj} = \delta^i_j. \tag{42}$$

The standard quadratic Casimir invariant is given in terms of this inverse as follows:

$$C = \frac{1}{2}g^{CK}_{ij}z^i z^j. \tag{43}$$

Journal of Physics: Conference Series **169** (2009) 012006 doi:10.1088/1742-6596/169/1/012006

If the Lie algebra is compact as well as semisimple then g_{CK} generates a gradient flow. In this case there are no degeneracies in g_{CK} and consequently the system relaxes without preserving an H nor an S.

Darboux's theorem of geometric mechanics reveals that all symplectic manifolds are equivalent in the sense that they all can be described locally in terms of canonical coordinate systems. In this vein, there is a similar structure theorem for metriplectic manifolds and double bracket dissipation with a Lie-Poisson bracket and Cartan-Killing dissipation. In the classification of Lie algebras, it is known that a basis exists such that the CK metric can be written in standard form where $g_{CK} = dia(\epsilon_1, \epsilon_2, \dots,)$ with $\epsilon_i \in \{-1, 0, 1\}$. A linear change of phase space coordinates, $\bar{z}^i = A^i_j z^j$ induces a basis change for the Lie algebra, which amounts to a usual tensorial change of coordinates for the structure constants and the Cartan-Killing metric. Thus, the work of Cartan and others can be carried over directly to give a structure theorem.

To illustrate some of the ideas above, consider the Lie algebras of dimension three [28]. The designation used here will be the same as that used for the homogeneous Bianchi cosmologies [29]. There are nine real three-dimensional Lie algebras, but only Types III, VIII, and IX will be considered for the purpose of demonstration. The structure constants, for these three cases are as follows:

(i) Type IX $\quad c_k^{ij} = \epsilon_{ijk}$,

(ii) Type VIII $\quad c_1^{23} = -c_1^{32} = -1\,, c_2^{31} = -c_2^{13} = 1\,, c_3^{12} = -c_3^{21} = 1$, otherwise 0,

(iii) Type III $\quad c_1^{13} = -c_1^{31} = +1$ otherwise 0.

Type IX is $so(3)$ and Type VIII is $sl(2,1)$, which arises in the context of a vortex reduction of the Hamiltonian description of fluid mechanics, e.g. the Kida vortex [30, 31]. The cosymplectic forms for the Lie-Poisson brackets for these algebras are

$$J_{IX} = \begin{pmatrix} 0 & z^3 & -z^2 \\ -z^3 & 0 & z^1 \\ z^2 & -z^1 & 0 \end{pmatrix}, \tag{44}$$

$$J_{VIII} = \begin{pmatrix} 0 & -z^3 & -z^2 \\ z^3 & 0 & z^1 \\ z^2 & -z^1 & 0 \end{pmatrix}, \tag{45}$$

$$J_{III} = \begin{pmatrix} 0 & 0 & z^1 \\ 0 & 0 & 0 \\ -z^1 & 0 & 0 \end{pmatrix}, \tag{46}$$

which have the Casimir invariants

$$C_{IX} = (z^1)^2 + (z^2)^2 + (z^3)^2\,, \tag{47}$$

$$C_{VII} = (z^1)^2 + (z^2)^2 - (z^3)^2\,, \tag{48}$$

$$C_{III} = f(z^1, z^2)\,, \tag{49}$$

respectively. The Cartan-Killing metrics, g_{CK}, for these algebras are

$$g_{IX} = -2\begin{pmatrix} 1 & 0 & 0 \\ 0 & 1 & 0 \\ 0 & 0 & 1 \end{pmatrix}, \tag{50}$$

$$g_{VIII} = -2\begin{pmatrix} -1 & 0 & 0 \\ 0 & 1 & 0 \\ 0 & 0 & 1 \end{pmatrix}, \tag{51}$$

$$g_{III} = -\begin{pmatrix} 0 & 0 & 0 \\ 0 & 0 & 0 \\ 0 & 0 & 1 \end{pmatrix}. \tag{52}$$

Note, Type IX is compact and semisimple, Type VIII is semisimple and noncompact, while Type III is not semisimple. To within a scale factor these are already in the standard form described above.

Because g_{IX} is the Euclidean metric, Type IX metriplectic dynamics and double bracket dynamics are exactly as before. It is, however, somewhat interesting to see what kind of dissipation is generated by the standard Casimir invariant of (43) when it is inserted into $(f,g)_{CK} = (f,g)_{IX}$, the dissipative bracket for this case. This gives

$$(z^i, C_{IX})_{IX} = g^{ij}_{IX} g^{IX}_{jk} z^k = z^i\,, \tag{53}$$

which corresponds to isotropic linear damping or growth. Note, this holds true for any compact semisimple Lie Group of any dimension.

The lack of definiteness for Type IIV is problematic if one seeks relaxation. However, if one chooses an S independent of z^1, then partial relaxation to the $z^2 - z^3$ plane is possible. Or, conversely, one could choose S independent of z^2 and z^3 and consider a relaxation to the z^1 axis. If such an S does not appear to have a mathematical or physical origin, then this is not too appealing.

Because of the degeneracy, Type III is more interesting. From (52) and (49), it is seen that $\partial C_{III}/\partial z$ is a null eigenvector of g_{III}. Consequently, the dynamics

$$\dot{z}^i = [z^i, F]_{III} + (z^i, F)_{III}\,, \tag{54}$$

with $F = H + C_{III}$, satisfies $\dot{C} = 0$, $\dot{F} = \dot{H}$ and

$$\dot{H} = (H,H)_{III} \leq 0\,. \tag{55}$$

Thus, this is an alternative to double bracket dynamics for incomplete systems.

One could continue and construct metriplectic dynamics for complete systems by projection etc., but this will not be pursued further now.

5. Summary and Conclusions

In this work, a variety of bracket formulations of dynamical systems have been described, with a division into the two categories of incomplete and complete systems. Incomplete systems do not describe in entirety the dynamics of both entropy and energy, but are quite useful for practical situations. Complete systems, as described by metriplectic dynamics, are a dynamical extension of thermodynamics that describes energy conservation and entropy production dynamics. An new kind of dissipation based on the Cartan-Killing metric was introduced and described for examples associated with Lie algebras of dimension three. The results about Cartan-Killing dissipation are preliminary: only the surface has been scratched and there are many directions in which one could proceed. This might be the subject of a future publication.

Journal of Physics: Conference Series **169** (2009) 012006 doi:10.1088/1742-6596/169/1/012006

Appendix A. Reminiscence

For many decades Allan Kaufman has been a most important presence in plasma physics. He has a widespread reputation for crisp, clear, and beautiful physics at the forefront of plasma science, with a theoretical physics flavor that belies his training at the University of Chicago under Gregor Wentzel. Allan has also had a profound influence by advising generations of students and by encouraging other young physicists. I personally owe him a debt of gratitude and I am glad to have this opportunity to express it and to reminisce a bit.

In the Fall of 1979 I moved to Princeton to accept a post doctoral position and while there I was most fortunate to eventually work with John Greene, another giant of our field. But, this did not happen right away. After a month or two at Princeton, I summoned the courage to poke my head into John's office and utter something that had been on my mind for a while, namely, that I thought MHD was a Hamiltonian theory. I am not exactly sure of the origin of this idea, but I had studied Hamiltonian dynamics as a graduate student and knew that this Berkeley plasma physics professor by the name of Kaufman had an interest in it and its applications to plasma physics. I remember in 1973 I wanted to know more about and possibly even be part of Allan's Berkeley school. In any case, my thought about MHD was met by John's 'go-away' stare – and so I left. Nevertheless, now and then I began toiling away on my own on this idea, totally ignorant of any literature on the subject. A while later, I returned to John and repeated that I thought MHD was a Hamiltonian theory – but added, here is why. John was more receptive this time and let me explain that based on a scaling argument the velocity field could not be a canonical field and that one needed to add a gradient and that this could be seen in Fourier space etc. John's response was, "Oh That." It turned out that I was well on my way to rediscovering the Clebsch representation of fluid mechanics, a potential representation of the velocity field dating from the mid-nineteenth century (see e.g. [22]). This was a big disappointment to me, but as things usually go, not a total loss. John suggested some literature, one thing led to another, and eventually this led to our fairly influential paper [1] on noncanonical brackets for MHD that was submitted to PRL in April of 1980. My Clebsch material was used as one of two means for proving the Jacobi identity by exploiting the connection between the canonical Clebsch variables and the usual but noncanonical Eulerian variables, which John insisted we do prior to publication. I wanted to give some backstory about how this came to pass, since many people have asked me about it and because Allan played a role in this story.

In January of 1980 there was a buzz around the theory wing of the Princeton Plasma Physics Laboratory that one of Allan's students was coming to give a seminar. Allan's students were (and are) viewed as gems: smart, highly polished, articulate, and universally well-educated. Robert Littlejohn came to Princeton and didn't disappoint when he spoke about his dissertation work on guiding center perturbation theory using noncanonical variables [32] that he did under Allan's supervision. Robert gave a sparkling and highly motivating talk. His talk and one of the papers that John suggested to me by Clifford Gardner on the Hamilton description of the KdV equation [24] were the inspiration for our work. Both concern Hamiltonian structure in terms of Poisson brackets that obey the Lie algebraic properties, but the latter was for an infinite-dimensional system and possesses a degeneracy and, hence, what we now call a Casimir invariant. This served as the prototype of what we called the noncanonical Hamiltonian formulation in our paper.

Allan immediately took an active interest in this work and gave me more encouragement than anyone. Many of us early in our careers have doubts and insecurities, and having an esteemed scientist like Allan take an interest was flattering, motivating, and also somewhat overwhelming. Allan's efforts were most important to me, because my style of mathematical physics was uncommon in the plasma community and my paper was met with criticism. Without his help, I may not have continued. Allan also played an active role in the dissemination of this work. Through him, Jerry Marsden and Alan Weinstein became aware of it, and this started their interest in noncanonical brackets for fluids and plasmas, followed by many others. Without question, because of Allan, my work rapidly cut a much wider swath.

Finally, I would like to mention another debt I owe Allan. Before I left Princeton he nominated me for a fellowship; although I didn't receive it, his action had consequences. First, he sent me a copy of the nomination letter, which was quite flattering. I remember carrying it around, rereading it, unbelieving that he had written such words. Second, he sent it to several senior scientists, and it was evident to me that their attitudes toward me changed because of this. I believe this action of Allan's jump-started my career. Last, at some point this letter got sent to my mother-in-law, an outspoken Cal graduate, and I do believe her attitude toward me also improved after that. So, Allan, thanks again for all you've done, for being a generous and kind mentor, and I wish you the happiest of Happy Birthdays.

Acknowledgment
This work was supported by U.S. Dept. of Energy Contract No. DE-FG05-80ET-53088.

References

[1] Morrison P J and Greene J M 1980 *Phys. Rev. Lett.* **45** 790–793
[2] Morrison P J 1980 *Phys. Lett. A* **80** 383–386
[3] Kaufman A N and Morrison P J 1982 *Phys. Lett. A* **88** 405–406
[4] Kaufman A N 1984 *Phys. Lett. A* **100** 419–422
[5] Morrison P J 1984 *Phys. Lett. A* **100** 423–427
[6] Morrison P J 1984 *Center for Pure and Applied Mathematics Report* **PAM-228** University of California at Berkeley
[7] Morrison P J 1986 *Physica D* **18** 410–419
[8] Turski L A and Kaufman A N 1987 *Phys. Lett. A* **120** 331–333
[9] Grmela M 1986 *Physica D* **21** 179–212
[10] Beris A N and Edwards B J 1990 *J. Rheol.* **34** 55–78
[11] Holm D, Putkaradze V and Tronci C 2008 *J. Phys. A* **41** 344010
[12] Brockett R W 1988 *Proc. IEEE* **27** 799
[13] Vallis G K, Carnevale G and Young W R 1989 *J. Fluid Mech.* **207** 133–152
[14] Morrison P J 1982 *Mathematical Methods in Hydrodynamics and Integrability in Dynamical Systems* vol 88 ed Tabor M and Treve Y (New York: Am. Inst. Phys.) pp 13–46
[15] Morrison P J 1998 *Rev. Mod. Phys.* **70** 467–521
[16] Marsden J E and Ratiu T S 1999 *Introduction to Mechanics and Symmetry* (New York: Springer)
[17] Morrison P J 2005 *Phys. Plasmas* **12** 058102–1–13
[18] Bloch A M 2003 *Nonholonomic Mechanics and Control* (New York: Springer)
[19] Otto F 2001 *Com. Partial Diff. Eqs.* **26** 101–174
[20] Hamilton R 1982 *J. Diff. Geom.* **17** 255–306
[21] Chow B and Knopf D 2004 *The Ricci Flow: An Introduction* (Providence: American Mathematical Society.)
[22] Serrin J 1959 *Handbuch der Physik* vol VIII, pt. 1 ed Flügge S (Berlin: Springer) pp 25–263
[23] Morrison P J 1985 *Talk presented at Kruskal festshrift, unpublished* (Institute for Theoretical Physics, Santa Barbara, Ca)
[24] Gardner C S 1971 *J. Math. Phys.* **12** 1548–1551
[25] Bloch A M, Flaschka H and Ratiu T 1990 *Duke Math.* **61** 41–65
[26] Padhye N and Morrison P J 1996 *Plasma Phys. Reports* **22** 869–877
[27] Morrison P J 1987 *Zeitschrift f. Naturforschung* **42a** 115–123
[28] Jacobson N 1962 *Lie Algebras* (New York: Interscience)
[29] Ryan M P and Shepley L C 1975 *Homogeneous Relativistic Cosmologies* (Princeton: Princeton University Press)
[30] Ngan K, Meacham S and Morrison P J 1995 *Phys. Fluids* **4** 896
[31] Meacham S P, Morrison P J and Flierl G R 1997 *Phys. Fluids* **9** 2310–2328
[32] Littlejohn R G 1979 *J. Math. Phys.* **20** 2445–2458

Journal of Physics: Conference Series **169** (2009) 012007 doi:10.1088/1742-6596/169/1/012007

Warm plasma model of intense, short-pulse laser-plasma interactions

Carl B. Schroeder

Lawrence Berkeley National Laboratory, Berkeley, California, 94720, USA

E-mail: CBSchroeder@lbl.gov

Abstract. A relativistic, warm fluid model of a nonequilibrium, collisionless plasma is developed. This model can be applied to describe nonlinear Langmuir waves excited by relativistically-intense, short-pulse lasers. Closure of the covariant fluid theory is obtained via an asymptotic expansion assuming a nonrelativistic plasma temperature. The quasi-static evolution equation for a laser-driven electron plasma wave is derived.

1. Introduction

Short-pulse (sub-ps), intense laser-plasma interactions access a physical regime where the plasma electrons experience relativistic motion while the plasma temperature (momentum spread) remains small. The plasma is typically created by the laser through photoionization and the laser-plasma interaction occurs on a time-scale short compared to the ion motion and the collision frequency. Such a collisionless plasma is not in local thermodynamical equilibrium, where standard relativistic collisionally-dominated plasma fluid theories apply (see, for example, Ref. [1]). In this paper, a relativistic warm plasma fluid theory of the nonequilibrium, collisionless plasma is described. This model can be used to examine the nonlinear plasma wave excitation by an intense laser pulse [2].

The electron plasma wave driven by an intense laser pulse can be highly-nonlinear and support large electric fields. Large amplitude plasma waves with relativistic phase velocities can be used to efficiently accelerate charged particles, and high-intensity lasers have been proposed for the excitation of relativistic plasma waves for plasma-based accelerator applications (for a review, see Ref. [3]). Laser-plasma-based accelerator experiments [4, 5, 6, 7] have measured electric fields in excess of 100 GV/m, several orders of magnitude greater then conventional accelerators. The relativistic warm fluid theory of the plasma described in this work can be applied to address the fundamental question of the maximum amplitude of these nonlinear relativistic electron plasma waves [2]. The maximum amplitude of the plasma wave sets the limit of the achievable accelerating field, and therefore, a fundamental limit to the development of plasma-based accelerator technology.

This paper is organized as follows. Section 2 presents a covariant fluid model of the collisionless plasma. Section 3 describes the warm plasma assumption that allows closure of the fluid equations. The quasi-static plasma response, assuming a travelling wave solution, is described in Sec. 4. Conclusions are offered in Sec. 5.

Journal of Physics: Conference Series **169** (2009) 012007 doi:10.1088/1742-6596/169/1/012007

2. Relativistic fluid equations

The collisionless Boltzmann equation [1] for particles of charge q and mass m is

$$\left[\frac{\partial}{\partial ct} + (\boldsymbol{p}/\gamma)\cdot\nabla + \frac{q}{mc^2}\left(\boldsymbol{E} + (\boldsymbol{p}/\gamma)\times\boldsymbol{B}\right)\frac{\partial}{\partial \boldsymbol{p}}\right] f = 0, \tag{1}$$

where $f(\boldsymbol{x},\boldsymbol{p},t)$ is the phase-space distribution (a Lorentz scalar), $\boldsymbol{p} = \gamma\boldsymbol{\beta}$ is the normalized particle momentum, and $\gamma^2 = 1 + p^2$. The collisionless Boltzmann equation can be written in the covariant form

$$p^\mu \partial_\mu f + \left(\frac{q}{mc^2}F^{\alpha\nu}p_\nu\right)\frac{\partial f}{\partial p^\alpha} = 0, \tag{2}$$

where $x^\mu = (ct, \boldsymbol{x})$, $p^\nu = (\gamma, \gamma\boldsymbol{\beta})$ is the normalized particle 4-momentum, $\partial^\mu = (\partial_{ct}, -\nabla)$, and $F^{\mu\nu} = \partial^\mu A^\nu - \partial^\nu A^\mu$ is the antisymmetric electromagnetic field-strength tensor [8], with $A^\mu = (\Phi, \boldsymbol{A})$ the 4-vector potential. The space-time metric tensor $g^{\mu\nu}$ for the following covariant formulation has the convention $g^{\mu\nu} = \mathrm{diag}(1, -1, -1, -1)$.

Following previous relativistic fluid formulations [9, 10, 11, 12], consider the following moments of the phase-space distribution:

$$h = \int d\Omega f, \tag{3}$$

$$J^\mu = \int d\Omega f p^\mu, \tag{4}$$

$$T^{\mu\nu} = \int d\Omega f p^\mu p^\nu, \tag{5}$$

$$M^{\alpha\mu\nu} = \int d\Omega f p^\alpha p^\mu p^\nu, \tag{6}$$

$$L^{\alpha\beta\mu\nu} = \int d\Omega f p^\alpha p^\beta p^\mu p^\nu, \tag{7}$$

where $d\Omega = d^3\mathbf{p}/p^0 = dp^1 dp^2 dp^3/p^0$ is the invariant momentum space volume element. Equations (3)–(6) correspond to the invariant density, fluid 4-current, energy-momentum stress tensor, and energy-momentum stress flux, respectively. The inhomogeneous Maxwell equations can be expressed as

$$\partial_\mu F^{\mu\nu} = 4\pi \sum_s q_s J_s^\nu, \tag{8}$$

where the sum is over species. The proper density may be defined using the 4-current as $n_p = (J^\mu J_\mu)^{1/2}$. Using the collisionless Boltzmann equation Eq. (2), the fluid description of the plasma (i.e., the evolution of the moments of the phase-space distribution) can be written in terms of the exact collisionless conservation laws:

$$\partial_\mu J^\mu = 0, \tag{9}$$

$$\partial_\mu T^{\mu\nu} = \left(\frac{q}{mc^2}\right) F^{\nu\sigma} J_\sigma, \tag{10}$$

$$\partial_\alpha M^{\alpha\mu\nu} = \left(\frac{q}{mc^2}\right) 2F^{(\mu\sigma}T_\sigma^{\nu)}, \tag{11}$$

$$\partial_\alpha L^{\alpha\beta\mu\nu} = \left(\frac{q}{mc^2}\right) 3F^{(\mu\sigma}M_\sigma^{\alpha\nu)}. \tag{12}$$

Equations (9)–(11) correspond to the fluid continuity (mass conservation), energy-momentum conservation, and conservation of energy-momentum flux, respectively. The following notation is used for indices based on the symmetrization convention: $K^{(\mu\nu)} = (K^{\mu\nu} + K^{\nu\mu})/2$, $K^{(\alpha\mu\nu)} = (K^{\alpha(\mu\nu)} + K^{\nu(\alpha\mu)} + K^{\mu(\alpha\nu)})/3$, and $K^{(\alpha\beta\mu\nu)} = (K^{\alpha(\beta\mu\nu)} + K^{\nu(\alpha\beta\mu)} + K^{\mu(\alpha\beta\nu)} + K^{\beta(\alpha\mu\nu)})/4$, where K is a tensor of some order.

Consider the normalized fluid 4-momentum, defined as the ratio of the 4-current to the invariant density,

$$u^\mu = J^\mu/h. \tag{13}$$

Using the fluid 4-momentum, the following centered moments of the phase-space distribution may be defined as

$$\Theta^{\mu\nu} = \int d\Omega f (p^\mu - u^\mu)(p^\nu - u^\nu), \tag{14}$$

$$Q^{\alpha\mu\nu} = \int d\Omega f (p^\alpha - u^\alpha)(p^\mu - u^\mu)(p^\nu - u^\nu), \tag{15}$$

$$R^{\alpha\beta\mu\nu} = \int d\Omega f (p^\alpha - u^\alpha)(p^\beta - u^\beta)(p^\mu - u^\mu)(p^\nu - u^\nu). \tag{16}$$

Note that there is freedom in choosing a fluid 4-momentum and various definitions may be employed (see, for example, Refs. [1, 13]). The fluid 4-momentum Eq. (13) is related to the more standard choice of Eckart [14] $U^\mu = J^\mu/(J_\mu J^\mu)^{1/2}$ by a Lorentz invariant scalar, namely the ratio of proper to invariant densities, $u^\mu = (n_p/h)U^\mu$.

The moments Eqs. (5)–(7) may be expressed using the centered moments Eqs. (14)–(16) as

$$T^{\mu\nu} = hu^\mu u^\nu + \Theta^{\mu\nu}, \tag{17}$$

$$M^{\alpha\mu\nu} = hu^\alpha u^\mu u^\nu + Q^{\alpha\mu\nu} + 3\Theta^{(\mu\nu}u^{\alpha)}, \tag{18}$$

$$L^{\alpha\beta\mu\nu} = hu^\alpha u^\beta u^\mu u^\nu + R^{\alpha\beta\mu\nu} + 6\Theta^{(\mu\nu}u^\alpha u^{\beta)} + 2u^{(\alpha}Q^{\beta)\mu\nu} + 2u^{(\mu}Q^{\nu)\alpha\beta}. \tag{19}$$

The contractions of the moments satisfy

$$g_{\mu\nu}T^{\mu\nu} = h, \tag{20}$$

$$g_{\mu\nu}M^{\alpha\mu\nu} = J^\alpha, \tag{21}$$

$$g_{\mu\nu}L^{\alpha\beta\mu\nu} = T^{\alpha\beta}, \tag{22}$$

and the contractions of the centered moments satisfy

$$g_{\mu\nu}\Theta^{\mu\nu} = (1 - u_\mu u^\mu)\, h, \tag{23}$$

$$g_{\mu\nu}Q^{\alpha\mu\nu} = -2\Theta^{\mu\alpha}u_\mu, \tag{24}$$

$$g_{\mu\nu}R^{\alpha\beta\mu\nu} = (1 - u_\mu u^\mu)\,\Theta^{\alpha\beta} - 2Q^{\mu\alpha\beta}u_\mu. \tag{25}$$

The conservation laws Eqs. (10)–(12) may be expressed using the centered moments as the fluid equations:

$$\partial_\mu(hu^\mu) = 0, \tag{26}$$

$$hu^\mu\partial_\mu u^\nu + \partial_\mu\Theta^{\mu\nu} = \left(\frac{q}{mc^2}\right) F^{\nu\alpha}hu_\alpha, \tag{27}$$

$$hu^\alpha\partial_\alpha(h^{-1}\Theta^{\mu\nu}) + 2\Theta^{(\mu\alpha}\partial_\alpha u^{\nu)} + \partial_\alpha Q^{\alpha\mu\nu} = \left(\frac{q}{mc^2}\right) 2F^{(\nu\alpha}\Theta^{\mu)}_\alpha, \tag{28}$$

$$hu^\alpha\partial_\alpha(h^{-1}Q^{\beta\mu\nu}) + 3Q^{\alpha(\mu\nu}\partial_\alpha u^{\beta)} + \partial_\alpha R^{\alpha\beta\mu\nu} = \left(\frac{q}{mc^2}\right) 3F^{(\nu\alpha}M^{\beta\mu)}_\alpha + 3h^{-1}\Theta^{(\mu\nu}\partial_\alpha\Theta^{\alpha\beta)}. \tag{29}$$

The above fluid formulation is an exact consequence of the collisionless Boltzmann equation Eq. (2) and no assumptions were made on the derivations of the above relations (the relativistic moment equations up to the fourth moment). In Appendix A the relativistic fluid quantities described in this section are related to conventional fluid quantities. In Sec. 3 a "warm" plasma assumption is considered, where it is assume that the distribution f has a small width about a relativistic mean momentum [13]. In particular, the third and higher-order centered moments (i.e., $Q^{\alpha\mu\nu}$, $R^{\alpha\beta\mu\nu}, \ldots$) are assumed to be small and may be neglected, providing closure to the hierarchy of moment equations.

3. Warm plasma assumption

In the following it is assumed that at any x^{μ}, the distribution f has a small momentum spread about its mean [13]. This warm plasma assumption will allow the hierarchy of moment equations to be truncated, and no specific form of the distribution f will be imposed.

An invariant measure of thermal spread may be defined as

$$\epsilon^2 = -\Theta^{\mu}_{\mu}/h = u^{\mu}u_{\mu} - 1, \tag{30}$$

where $|\epsilon| \ll 1$ is assumed to be small. Note that ϵ is a Lorentz invariant and allows a truncation valid in the rest frame to be preserved after an arbitrary Lorentz boost. In the local plasma rest frame (i.e., moving at u^{μ}), it is assumed that $\Theta^{\mu\nu}/h = \mathcal{O}(\epsilon^2)$ and $Q^{\alpha\mu\nu}/h = \mathcal{O}(\epsilon^3)$, and the components satisfy $\Theta^{ij}/h = \mathcal{O}\left(\epsilon^2\right)$, $\Theta^{i0}/h = \mathcal{O}\left(\epsilon^3\right)$, $\Theta^{00}/h = \mathcal{O}\left(\epsilon^4\right)$, and $Q^{ijk}/h = \mathcal{O}\left(\epsilon^3\right)$, $Q^{ij0}/h = \mathcal{O}\left(\epsilon^4\right)$, $Q^{i00}/h = \mathcal{O}\left(\epsilon^5\right)$, $Q^{000}/h = \mathcal{O}\left(\epsilon^6\right)$. Truncation of the moment hierarchy to order $\mathcal{O}(\epsilon^2)$ is achieved by neglecting the third-order centered moment $Q^{\alpha\mu\nu}$ in the fluid equations [Eq. (28)]. The invariant measure of thermal spread is related to the width of the distribution (i.e., temperature). In Appendix B, the invariant measure ϵ is calculated for a relativistic Maxwellian (Jüttner) distribution and shown to be equal to the width of the momentum distribution.

As a consequence of the thermal inertia, the invariant density h is not equal to the proper density defined by the 4-current $n_p = (J^{\mu}J_{\mu})^{1/2}$. Following Refs. [11, 9], consider the ratio $\lambda = n_p/h$. Note that $u^{\mu}u_{\mu} = \lambda^2$. Using the contraction of the stress energy tensor Eq. (20),

$$\lambda^2 = 1 - \Theta^{\mu}_{\mu}/h = 1 + \epsilon^2. \tag{31}$$

The normalized thermal velocity spread may be defined as

$$\beta^2_{\text{th}} = \frac{-\Theta^{\mu}_{\mu}}{h\lambda^2} = \lambda^{-2}\epsilon^2. \tag{32}$$

Equation (30) implies that $\epsilon^2 \ll 1$ if the rest frame thermal velocity of the plasma is nonrelativistic $\beta^2_{\text{th}} \ll 1$. With the definition Eq. (32), Eq. (31) can be expressed as

$$\lambda = (1 - \beta^2_{\text{th}})^{-1/2}, \tag{33}$$

and can be identified as the relativistic Lorentz factor associated with the thermal fluctuations. To order $\mathcal{O}(\epsilon^2)$, i.e., the warm plasma approximation, $\beta^2_{\text{th}} = \epsilon^2$ and $\lambda = 1 + \beta^2_{\text{th}}/2$. The fluid 3-velocity $\boldsymbol{w}$ may be defined such that $u^{\mu} = J^{\mu}/h = (\lambda\Gamma, \lambda\Gamma\boldsymbol{w})$, where $\Gamma^{-2} = (1 - \boldsymbol{w}\cdot\boldsymbol{w})$.

Neglecting the third-order centered moment $Q^{\alpha\mu\nu}$ [i.e., retaining terms to order $\mathcal{O}(\epsilon^2)$], the contraction Eq. (24) becomes

$$\Theta^{\mu\nu}u_{\mu} = 0, \tag{34}$$

Journal of Physics: Conference Series **169** (2009) 012007 doi:10.1088/1742-6596/169/1/012007

and the fluid equations Eqs. (26), (27), and (28) become [to order $\mathcal{O}(\epsilon^2)$]

$$u^\mu \partial_\mu h = -h\partial_\mu u^\mu, \tag{35}$$

$$u^\mu \partial_\mu u^\nu = -\partial_\mu \Theta^{\mu\nu} + (q/mc^2)F^{\nu\alpha}hu_\alpha, \tag{36}$$

$$u^\alpha \partial_\alpha \left(\Theta^{\mu\nu}/h\right) = -\Theta^{\mu\alpha}\partial_\alpha u^\nu - \Theta^{\nu\alpha}\partial_\alpha u^\mu + (q/mc^2)\left(F^{\nu\alpha}\Theta^\mu_\alpha + F^{\mu\alpha}\Theta^\nu_\alpha\right). \tag{37}$$

Equations (34)–(37) constitute a closed set of fluid equations, which may be solved for the plasma evolution.

3.1. Relativistic Bohm-Gross dispersion relation

Applying the Fourier transformation $(z, t) \to (k, \omega)$ to the linearized relativistic warm fluid equations [i.e., linearization of Eqs. (34)–(37)], coupled to the Poisson equation for an electrostatic space-charge oscillation, yields the Langmuir wave dispersion relation:

$$\omega^2 = \omega_p^2(1 - 5\theta/2) + 3\theta k^2 c^2, \tag{38}$$

where θ is the initial isotropic plasma temperature (invariant measure of thermal spread). The dispersion relation Eq. (38) was first derived by Clemmow and Wilson [15] and is the generalization of the Bohm-Gross dispersion relation [16] to include thermal inertia (enthalpy). The linear wavelength is

$$\lambda \simeq \lambda_p \beta_\varphi \left(1 + \frac{5}{4}\theta - \frac{3}{2}\frac{\theta}{\beta_\varphi^2}\right), \tag{39}$$

where $\beta_\varphi = \omega/(kc)$ is the phase velocity of the plasma wave and $\lambda_p = 2\pi c/\omega_p$. For linear plasma waves with relativistic phase velocities, $\beta_\varphi \simeq 1$, the wavelength is $\lambda \simeq \lambda_p(1 - \theta/4)$.

3.2. Warm-plasma closure in the one-dimensional limit

In the 1D limit $u^\mu = (\lambda\Gamma, \lambda\Gamma w)$, and Eq. (34) implies $\Theta^{00} = w^2\Theta^{11}$ and $\Theta^{10} = w\Theta^{11}$. Combining the evolution equation for Θ^{11}, Eq. (37), the continuity equation Eq. (35), and the zeroth-component of the momentum equation Eq. (36), yields [to order $\mathcal{O}(\epsilon^2)$],

$$(\partial_0 + w\partial_1)\left(h^{-3}\Gamma^{-2}\Theta^{11}\right) = 0, \tag{40}$$

where $(\partial_0 + w\partial_1)$ is the (noncovariant) convective time derivative. Equation (40) is a result of entropy conservation. Assuming an initially quiescent ($w = 0$) plasma, Eq. (40) integrates to yield

$$\frac{\Theta^{11}}{n_0} = \Gamma^2\left(\frac{h}{n_0}\right)^3\theta, \tag{41}$$

where θ is the initial plasma temperature normalized to mc^2/k_{B}. In the one-dimensional limit, to order $\mathcal{O}(\epsilon^2)$, the temperature (invariant measure of thermal spread) is given by

$$\epsilon^2 = \frac{n_0}{h}\left(\frac{\Theta^{11}}{n_0} - \frac{\Theta^{00}}{n_0}\right) = \left(\frac{h}{n_0}\right)^2\theta. \tag{42}$$

Equation (41) provides closure to the fluid equations, and the momentum variance is proportional to the cube of the density. This is sometimes referred to as the "adiabatic equation of state" [17]. Note that, the derivation of Eqs. (41) and (42) does not assume a specific form of the phase-space distribution f, or an adiabatic equation of state (although only reversible processes are considered in this collisionless formulation). The derivation only assumes the asymptotic expansion (warm plasma approximation) is valid. Equation (41) is valid, in the one-dimensional limit, for any distribution f provided $\epsilon^2 = \theta(h/n_0)^2 \ll 1$ (i.e., a warm plasma), and the plasma is initially quiescent.

4. Quasi-static warm fluid equations

The warm fluid model can be applied to describe the response of the plasma to an intense laser field. It is assumed that the plasma wave and drive (e.g., laser field) are only functions of the co-moving variable $\xi = z - \beta_\varphi ct$ [i.e., the quasi-static approximation (QSA)], where β_φ is the phase velocity of the plasma wave (approximately group velocity of the driver). In the QSA, the phase-space distribution takes the form $f(\zeta, p^\mu)$. Note that, in the QSA, $\partial_1 = \partial_z = \partial_\xi$ and $\partial_0 = \partial_{ct} = -\beta_\varphi \partial_\xi$. For simplicity, in the following, we consider the one-dimensional limit with no momentum variance in the transverse direction: $f = g(\zeta, p_z)\delta^2(p_\perp - a_\perp)$, where $a_\perp = eA_\perp/mc^2$ is the normalized vector potential of the laser field (in the Coulomb gauge).

In the QSA, the continuity equation Eq. (9), $\partial_\mu(hu^\mu) = 0$, becomes

$$\partial_\xi \left[h\lambda\Gamma\left(\beta_\varphi - w_z\right)\right] = 0, \tag{43}$$

or, for an initially quiescent plasma,

$$h/n_0 = \left[\lambda\Gamma(1 - \beta_\varphi^{-1} w_z)\right]^{-1}. \tag{44}$$

For the case of no momentum variance in the transverse direction, the transverse momentum equation becomes

$$\partial_\xi \left(\lambda\Gamma w_\perp - a_\perp\right) = 0, \tag{45}$$

which can be integrated to yield $\lambda\Gamma w_\perp = a_\perp$, or

$$w_\perp = \frac{a_\perp(1 - w_z^2)^{1/2}}{\left(\lambda^2 + a_\perp^2\right)^{1/2}} = \frac{a_\perp(1 - w_z^2)^{1/2}}{\left(\gamma_\perp^2 + \epsilon^2\right)^{1/2}}. \tag{46}$$

In the QSA, for plasma wave excitation by a laser driver, the components of the energy-momentum conservation equation Eq. (10) are

$$\partial_\xi \left(-\beta_\varphi T^{00} + T^{10}\right) = h\lambda\Gamma\left(w_z \partial_\xi \phi + \beta_\varphi w_\perp \partial_\xi a_\perp\right), \tag{47}$$

$$\partial_\xi \left(-\beta_\varphi T^{01} + T^{11}\right) = h\lambda\Gamma\left(\partial_\xi \phi + w_\perp \partial_\xi a_\perp\right), \tag{48}$$

where $\phi = e\Phi/mc^2$ is the normalized electrostatic potential of the plasma wave. Combining Eqs. (47) and (48), and using the continuity equation Eq. (43), yields

$$\partial_\xi \left(-\beta_\varphi^2 T^{01} + \beta_\varphi T^{11} + \beta_\varphi T^{00} - T^{10}\right) = h\lambda\Gamma(\beta_\varphi - w_z)\partial_\xi \phi = n_0 \beta_\varphi \partial_\xi \phi. \tag{49}$$

Using $T^{\mu\nu} = hu^\mu u^\nu + \Theta^{\mu\nu}$, and Eq. (34) (i.e., $\Theta^{\mu 0} = w_z \Theta^{\mu 1}$), Eq. (49) becomes

$$\partial_\xi \left[\left(h\lambda^2\Gamma^2 + \Theta^{11}\right)(1 - \beta_\varphi w_z)(1 - \beta_\varphi^{-1} w_z)\right] = n_0 \partial_\xi \phi. \tag{50}$$

Equation (50) yields, to order $\mathcal{O}(\epsilon^2)$, the quasi-static longitudinal constant of motion for the distribution $f(\xi, p^\mu)$, assuming a plane laser field in an initially isotropic plasma:

$$\partial_\xi \left[\frac{(1 - \beta_\varphi w_z)\gamma_\perp}{(1 - w_z^2)^{1/2}} + \frac{3}{2}\theta \frac{(1 - \beta_\varphi w_z)(1 - w_z^2)^{1/2}}{\gamma_\perp (1 - \beta_\varphi^{-1} w_z)^2} - \phi\right] = 0, \tag{51}$$

with $\gamma_\perp = 1 + a_\perp^2$. Assuming an initially quiescent plasma, Eq. (51) can be integrated to yield the longitudinal constant of motion:

$$\frac{(1 - \beta_\varphi w_z)\gamma_\perp}{(1 - w_z^2)^{1/2}} + \frac{3}{2}\theta \frac{(1 - \beta_\varphi w_z)(1 - w_z^2)^{1/2}}{\gamma_\perp (1 - \beta_\varphi^{-1} w_z)^2} - \phi = 1 + \frac{3}{2}\theta. \tag{52}$$

The right-hand side of Eq. (52) corresponds to the rest energy of the plasma and the energy contained in the initial thermal fluctuations. For $\theta = 0$, Eq. (52) reduces to the well-known quasi-static result [3] for plasma wave excitation by an ultra-short laser pulse in a cold plasma.

The evolution of the plasma wave potential is determined by the Poisson equation

$$k_p^{-2}\partial_\xi^2\phi = J^0/n_0 - 1 = \lambda\Gamma h/n_0 - 1, \tag{53}$$

where $k_p^2 = \omega_p^2/c^2$. Equation (52) can be combined with the continuity equation Eq. (44) and the Poisson equation Eq. (53) to yield the evolution equation for the plasma fluid momentum [2]:

$$\frac{\partial^2}{\partial k_p^2\xi^2}\left[\frac{\gamma_\perp(1-\beta_\varphi w_z)}{(1-w_z^2)^{1/2}} + \frac{3}{2}\theta\frac{(1-\beta_\varphi w_z)(1-w_z^2)^{1/2}}{\gamma_\perp(1-\beta_\varphi^{-1}w_z)^2}\right] = \frac{w_z}{\beta_\varphi - w_z}. \tag{54}$$

The excited plasma wave electric field is given by $\hat{E} = E_z/E_0 = -k_p^{-1}\partial_\xi\phi(w_z)$, where $\phi(w_z)$ is given by Eq. (52) and $E_0 = mc^2k_p/e$. In terms of the axial fluid velocity, the plasma density is $n/n_0 = w_z/(\beta_\varphi - w_z)$. Equation (54) is an ordinary differential equation, which may be solved numerically for an arbitrary laser field $a_\perp(\xi)$.

5. Conclusions and discussion

In this work, a relativistic warm fluid theory of a nonequilibrium, collisionless plasma was presented. This warm fluid theory can be applied to model intense, short-pulse laser-plasma interactions. The fluid equations were treated asymptotically, and closure was obtained by assuming a nonrelativistic temperature $\epsilon^2 \ll 1$ (no specific form of the phase-space distribution f or equation of state was assumed). The relativistic warm fluid theory was used to calculate the evolution equation for a laser-driven plasma wave assuming a travelling wave (i.e., the quasi-static approximation).

If the plasma temperature becomes relativistic, the asymptotic expansion will no longer be valid. For relativistic temperatures, the higher-order moments of the distribution will be important and will be a function of the specific form of the phase-space distribution. Note that, choice of an unphysical distribution (e.g., water-bag) may lead to singular (unbounded) solutions in the relativistic temperature regime. In addition, for sufficiently large density perturbations, the collisionless plasma model will no longer be valid and collisional effects will become important.

In addition to the warm plasm assumption, wave-particle interactions were neglected in this theory. This assumption will be valid provided the phase velocity of the plasma wave is larger than the thermal velocity spread $\beta_\varphi^2 \gg \beta_{\rm th}^2$, such that the bulk of the plasma distribution is far from resonance with the wave. As the field amplitude approaches and exceeds the wavebreaking limit (maximum traveling wave amplitude), particles in the tail of the plasma distribution can become trapped in the plasma wave [18, 19]. When sufficient number of electrons becomes trapped, the wave will be perturbed such that the quasi-static assumption will no longer be valid. Proper characterization of the phase-space distribution is critical to analyzing kinetic effects such as particle trapping in intense laser-plasma interactions.

As well as modeling intense laser-plasma interactions, the relativistic, warm fluid theory of a nonequilibrium, collisionless plasma presented here can also be applied to model relativistic charged particle beams and beam-plasma interactions, and may have applications to astrophysical plasmas.

Acknowledgments

I would like to express my great thanks to Prof. Allan Kaufman for teaching me the fundamental aspects of theoretical plasma physics while I was a graduate student at Berkeley, and to offer my best wishes on the occasion of his eightieth year.

I would also like to thank B. A. Shadwick, A. Charman, E. Esarey, and R. Lindberg for many enlightening discussions on the topic of laser-driven plasma waves.

This work was supported by the Director, Office of Science, Office of High Energy Physics, of the U.S. Department of Energy under Contract No. DE-AC02-05CH11231.

References

[1] de Groot S R, van Leeuwen W A and van Weert C G 1980 *Relativistic Kinetic Theory* (Amsterdam: North-Holland)
[2] Schroeder C B, Esarey E and Shadwick B A 2005 *Phys. Rev. E* **72** 055401
[3] Esarey E, Sprangle P, Krall J and Ting A 1996 *IEEE Trans. Plasma Sci.* **24** 252–288
[4] Mangles S P D, Murphy C D, Najmudin Z, Thomas A G R, Collier J L, Dangor A E, Divall E J, Foster P S, Gallacher J G, Hooker C J, Jaroszynski D A, Langley A J, Mori W B, Norreys P A, Tsung F S, Viskup R, Walton B R and Krushelnick K 2004 *Nature* **431** 535–538
[5] Geddes C G R, Toth C, van Tilborg J, Esarey E, Schroeder C B, Bruhwiler D, Nieter C, Cary J and Leemans W P 2004 *Nature* **431** 538–541
[6] Faure J, Glinec Y, Pukhov A, Kiselev S, Gordienko S, Lefebvre E, Rousseau J P, Burgy F and Malka V 2004 *Nature* **431** 541–544
[7] Leemans W P, Nagler B, Gonsalves A J, Tóth C, Nakamura K, Geddes C G R, Esarey E, Schroeder C B and Hooker S M 2006 *Nature Phys.* **2** 696–699
[8] Jackson J D 1975 *Classical Electrodynamics* (New York: Wiley)
[9] Newcomb W A 1982 *Phys. Fluids* **25** 846–851
[10] Amendt P 1986 *Phys. Fluids* **29** 1458–1471
[11] Siambis J G 1986 *Phys. Fluids* **30** 896–903
[12] Pennisi S and Anile A M 1990 *Phys. Fluids* **3** 1091–1103
[13] Shadwick B A, Tarkenton G M and Esarey E H 2004 *Phys. Rev. Lett.* **93** 175002
[14] Eckart C 1940 *Phys. Rev.* **58** 919–924
[15] Clemmow P C and Wilson A J 1956 *Proc. R. Soc. London Ser. A* **237** 117–131
[16] Bohm D and Gross E P 1949 *Phys. Rev.* **75** 1851–1864
[17] Kruer W L 1988 *The Physics of Laser Plasma Interactions* (Redwood City, CA: Addison-Wesley)
[18] Schroeder C B, Esarey E, Shadwick B A and Leemans W P 2006 *Phys. Plasmas* **13** 033103
[19] Esarey E, Schroeder C B, Cormier-Michel E, Shadwick B A, Geddes C G R and Leemans W P 2007 *Phys. Plasmas* **14** 056708
[20] Taub A H 1948 *Phys. Rev.* **74** 328–334
[21] Jüttner F 1911 *Ann. Physik* **34** 856

Appendix A. Relation between classical fluid quantities and warm fluid model

In this appendix, the relativistic fluid model quantities of Sec. 2 are related to conventional fluid quantities (see, e.g., Ref. [1]). The energy density may be defined as

$$\mathsf{e}n_p = U_\mu T^{\mu\nu} U_\nu, \tag{A.1}$$

and the pressure tensor may be defined as

$$P^{\mu\nu} = \Delta^\mu_\alpha T^{\alpha\beta} \Delta^\nu_\beta = \mathsf{p}\Delta^{\mu\nu} + \Pi^{\mu\nu}, \tag{A.2}$$

where p is the hydrostatic pressure, $\Delta^{\mu\nu} = g^{\mu\nu} - U^\mu U^\nu$, and $\Pi^{\mu\nu}$ is the irreversible or viscous part of the pressure tensor. If $\Pi^{\mu\nu} = 0$ (i.e., collisionless plasma), then the hydrostatic pressure is

$$\mathsf{p} = \left(\Delta^\mu_\mu\right)^{-1} \Delta_{\alpha\beta} T^{\alpha\beta}. \tag{A.3}$$

Using the contraction $T^\mu_\mu = h$, the invariant density (i.e., enthalpy) is

$$h = \mathsf{e}n_p + \Delta^\mu_\mu \mathsf{p}. \tag{A.4}$$

In one dimension, $\Delta^\mu_\mu = -1$, and, in 3D, $\Delta^\mu_\mu = -3$.

In the warm fluid model, the hydrostatic pressure is

$$\mathsf{p} = \left(\Delta^{\mu}_{\mu}\right)^{-1} g_{\mu\nu}\Theta^{\mu\nu} = -\left(\Delta^{\mu}_{\mu}\right)^{-1} h\epsilon^2. \tag{A.5}$$

The 1D hydrostatic pressure is given by $\mathsf{p} = h\epsilon^2$. Recall that $(n_p/h)^2 = 1+\epsilon^2$. With the warm plasma assumption, the energy density $\mathsf{e}n_p = U_\mu T^{\mu\nu} U_\nu$ is

$$\mathsf{e}n_p = h - \Delta^{\mu}_{\mu}\mathsf{p} = h(1+\epsilon^2). \tag{A.6}$$

The heat flow can be defined as $\mathsf{q}^\mu = U_\nu T^{\nu\alpha}\Delta^{\mu}_{\alpha}$ (difference of the energy flow and flow of enthalpy). Evaluating the heat flow yields,

$$\mathsf{q}^\mu = (h/n_p)\left[U_\sigma Q^{\alpha\sigma}_{\alpha} U^\mu - Q^{\mu\alpha}_{\alpha}\right]/2 = -\frac{1}{2}\frac{h}{n_p} Q^{\alpha\mu}_{\alpha} + \frac{1}{4}\left(\frac{h}{n_p}\right)^2 U^\mu \left[(\Theta^{\nu}_{\nu})^2/h - R^{\delta\alpha}_{\delta\alpha}\right]. \tag{A.7}$$

The heat flow is proportional to the third-order (and higher) moments, i.e., $q^\mu \sim h\epsilon^3$. The heat flow is only proportional to higher-order moments because a collisionless plasma (i.e., no viscosity) was assumed.

The Taub inequality [20], $\mathsf{e}h \geq n_p$, can be expressed in terms of the centered moments as $(\Theta^{\nu}_{\nu})^2 \leq hR^{\alpha\delta}_{\alpha\delta}$, which is the Cauchy-Schwarz inequality for the moments of the distribution. Within the warm plasma approximation, to $\mathcal{O}(\epsilon^2)$, $\mathsf{e}h = n_p$.

Appendix B. Invariant measure of thermal spread of a Jüttner distribution

In this appendix, the invariant measure of thermal spread ϵ is calculated assuming a Jüttner (relativistic Maxwell-Boltzmann) distribution. It is shown that the invariant measure of thermal spread is related to the width of the distribution.

Consider the Jüttner distribution [1, 21] given by

$$f = f_0 \exp\left[-p^\mu U_\mu/\theta_{\mathrm{J}}\right], \tag{B.1}$$

where θ_{J} is the equilibrium temperature of the system (normalized to mc^2/k_{B}) and f_0 is a constant. Note that, since $p^\mu U_\mu$ is a Lorentz scalar, $p^\mu U_\mu$ is invariant and may be evaluated in any frame. In the rest frame of the plasma fluid $U^\mu = (1, \mathbf{0})$ and $p^\mu U_\mu = \gamma$. In the rest frame of the plasma fluid, and assuming a warm plasma, $f = f_0\exp(-\gamma/\theta_{\mathrm{J}}) = f_0\exp[-(1-\beta^2)^{-1/2}/\theta_{\mathrm{J}}] \simeq f_1\exp(-\beta^2/2\theta_{\mathrm{J}})$, and the equilibrium temperature θ_{J} is the root-mean square of the plasma velocity distribution.

Appendix B.1. 1D distribution

In 1D, the invariant momentum-space volume is $d\Omega = dp/p^0 = dp/\gamma = \gamma d\gamma/(\gamma^2-1)^{1/2}$. For a Jüttner distribution Eq. (B.1) in 1D, the invariant density is

$$h = \int f d\Omega = f_0 \int_{-\infty}^{\infty} \frac{dp}{p^0} e^{-p^\mu U_\mu/\theta_{\mathrm{J}}} = 2f_0 \int_1^\infty \left(\gamma^2-1\right)^{-1/2} e^{-\gamma/\theta_{\mathrm{J}}} d\gamma = 2f_0 K_0(\theta_{\mathrm{J}}^{-1}), \tag{B.2}$$

and the proper density $n_p = U_\mu J^\mu$ is

$$n_p = \int U_\mu p^\mu f d\Omega = f_0 \int_{-\infty}^{\infty} \frac{dp}{p^0}(p^\mu U_\mu) e^{-(p^\mu U_\mu)/\theta_{\mathrm{J}}} = 2f_0 \int_1^\infty \frac{\gamma e^{-\gamma/\theta_{\mathrm{J}}}}{(\gamma^2-1)^{1/2}} d\gamma = 2f_0 K_1(\theta_{\mathrm{J}}^{-1}). \tag{B.3}$$

Journal of Physics: Conference Series **169** (2009) 012007 doi:10.1088/1742-6596/169/1/012007

The ratio of proper to invariant densities is $\lambda = n_p/h = K_1(\theta_J^{-1})/K_0(\theta_J^{-1})$, and the invariant measure of thermal spread is

$$\epsilon^2 = \left[K_1(\theta_J^{-1})/K_0(\theta_J^{-1})\right]^2 - 1. \tag{B.4}$$

Note, for large argument $A \gg 1$, $K_n(A) = \sqrt{(\pi/2)}A^{-1/2}e^{-A}[1 + (4n^2 - 1)/(8A) + \cdots]$. In the limit of nonrelativistic temperature $\theta_J \ll 1$,

$$\epsilon^2 = \frac{(1 + 3\theta_J/8 + \cdots)^2}{(1 - \theta_J/8 + \cdots)^2} - 1 \simeq \theta_J. \tag{B.5}$$

The invariant measure of thermal spread can be identified as the equilibrium temperature of a Jüttner (relativistic Maxwell-Boltzmann) distribution, $\epsilon^2 = \theta_J$.

Appendix B.2. 3D distribution

In 3D, the invariant momentum-space volume is $d\Omega = d^3\mathbf{p}/p^0 = 4\pi p^2 dp/\gamma = \gamma d\gamma(\gamma^2 - 1)^{1/2}$. For a Jüttner distribution Eq. (B.1) in 3D, the invariant density is

$$h = f_0 \int_{-\infty}^{\infty} \frac{d^3\mathbf{p}}{p^0} e^{-p^\mu U_\mu/\theta_J} = 4\pi f_0 \int_1^{\infty} \left(\gamma^2 - 1\right)^{1/2} e^{-\gamma/\theta_J} d\gamma = 4\pi f_0 \theta_J K_1(\theta_J^{-1}), \tag{B.6}$$

and the proper density $n_p = U_\mu J^\mu$ is

$$n_p = f_0 \int_{-\infty}^{\infty} \frac{d^3\mathbf{p}}{p^0} (p^\mu U_\mu) e^{-(p^\mu U_\mu)/\theta_J} = 4\pi f_0 \int_1^{\infty} \gamma \left(\gamma^2 - 1\right)^{1/2} e^{-\gamma/\theta_J} d\gamma = 4\pi f_0 \theta_J K_2(\theta_J^{-1}). \tag{B.7}$$

The ratio of proper to invariant densities is $\lambda = n_p/h = K_2(\theta_J^{-1})/K_1(\theta_J^{-1})$, and the invariant measure of thermal spread is

$$\epsilon^2 = \left[K_2(\theta_J^{-1})/K_1(\theta_J^{-1})\right]^2 - 1. \tag{B.8}$$

In the limit of nonrelativistic temperature $\theta_J \ll 1$,

$$\epsilon^2 = \frac{(1 + 15\theta_J/8 + \cdots)^2}{(1 + 3\theta_J/8 + \cdots)^2} - 1 \simeq 3\theta_J. \tag{B.9}$$

Journal of Physics: Conference Series **169** (2009) 012008 doi:10.1088/1742-6596/169/1/012008

Allan Kaufman's contributions to plasma wave theory

E. R. Tracy[1], **A. J. Brizard**[2]

[1] Department of Physics, College of William and Mary, Williamsburg, VA 23187-8795, USA
[2] Saint Michael's College, Colchester, VT 05439, USA

E-mail: ertrac@wm.edu

Abstract. A brief review is presented of the contributions of Allan Kaufman to the theory of plasma waves. These contributions have been rich and various, characterized by high quality and a sense of what makes a problem both interesting and important. We include a brief summary of work prior to the mid-1980's, but the primary emphasis will be on more recent work concerning the use of phase space methods in the theory of linear plasma waves and mode conversion. One goal of the paper is to place Allan's contributions in the wider context of semi-classical methods. We will emphasize the underlying intuitions rather than providing a detailed mathematical exposition which can be found in the literature cited.

1. Introduction

The authors have had the great privilege of working with Allan Kaufman for nearly twenty years and can attest to the fact that he combines deep physical intuition about plasmas, with a strong mix of geometrical and analytical skills. Both of the authors were attracted to work with Allan because of his track record of working on important physics problems, and that more elusive quality we usually call 'style'. In this paper we take the opportunity afforded by the KaufmanFest 07 conference proceedings to step back and provide a look at Allan's contributions in the area of plasma wave theory. Our goal is not to give a detailed technical discussion of his achievements in this area, which can be found in the papers cited, but to instead provide an overview of his many years of productive work. This will show how keen Allan's instincts have been for finding interesting problems, and to highlight the elegance of his approach to physics. Allan's work on the closely related areas of action principles, geometrical perturbation theory, and gyrokinetics, are described elsewhere in these proceedings.

Because this summary is intended to describe Allan's contributions in plasma wave theory, rather than providing a detailed review of this topic, we apologize in advance for researchers whose papers are not cited here.

Allan has made seminal contributions in the areas of wave chaos, wave kinetic equations, ponderomotive effects, quasilinear diffusion and mode conversion in nonuniform plasmas. For many of these topics, Allan was one of the first in the plasma community to contribute. For others, Allan's contributions are often considered to be among the clearest or most insightful. Further, he was often responsible for keeping plasma physics 'out front' among the other fields of physics, developing ideas and tools that have since been adopted by others, or where other fields have yet to catch up.

Although we will attempt to provide a reasonably full historical survey of Allan's most significant contributions in plasma wave theory, we assert the authors' prerogative to focus on areas that we are most familiar with. Hence, we will break the historical survey into two 'epochs': before, and after, the mid-1980's. Prior to the mid-80's, Allan's work was largely concerned with nonlinear effects, as was much of the plasma community. The authors can recall as graduate students learning the general opinion that linear theory was basically 'done' and all the 'real' problems were nonlinear. Hence, it was a bit of shock to find out this was not the case, and that there were still many problems in linear wave theory that demanded attention. Among these were problems in vector WKB theory, such as covariant formulations, wave-kinetic equations for incoherent fields, and how to deal with regions where WKB methods break down such as caustics, mode conversion and gyroresonance. By the end of the 1980's, Allan's work in linear wave theory took on the distinctive flavor of modern 'semi-classical' analysis, with its emphasis on ray phase-space methods.

Given how much of Allan's work in plasma wave theory has centered on the use of WKB methods, it is interesting to note that Allan received his PhD from the University of Chicago under the direction of Gregor Wentzel, the 'W' in WKB. Hence, Allan has a 'Wentzel number' of one. Allan's recollections of his work with Wentzel are described in his memoirs, also a part of this volume.

The outline of our paper is as follows: In Section 2 we provide a quick review of Allan's contributions to plasma wave theory prior to 1985. This break is somewhat arbitrary, but it marks a point where Allan's approach to the theory of waves began to incorporate and develop new ideas from semi-classical analysis. Hence, the work after this time becomes more informed by the notion of ray phase space. His research interests prior to this time are discussed in his memoirs, so we will pass over them quickly. One pattern that emerges from this survey is how intellectually ambitious Allan has been, and how open he is to new ideas and methods. As a result, Allan has often been among the first in plasma theory to exploit new mathematical ideas, or concepts and tools from other fields.

Much of Allan's work since 1985 has concerned WKB theory, and extensions of that theory that are required in various circumstances, so in Section 3 we provide an exceedingly brief summary of WKB methods and ray tracing. Particular emphasis is placed on new ideas that have been introduced since the 1960's.

In Section 4 we describe Allan's contributions, in collaboration with Steve McDonald, to the theory of 'quantum chaos' (although Allan's work was actually done in the context of classical wave equations) and a closely related derivation of the wave-kinetic equation for electromagnetic waves. The wave-kinetic equation makes the important assumption that there are no non-adiabatic (mode conversion or resonance crossing) effects, though it did allow for the presence of weak sources and dissipation. The vector character of the problem introduces significant complications that do not occur in the scalar wave problem, and the approach taken by Kaufman and McDonald introduces some key ideas that will reappear in much of Allan's subsequent work.

In Section 5 we describe what Lazar Friedland has called Allan's 'conversion to mode conversion'. In this section we stick to a configuration space viewpoint of conversion and draw heavily upon some notes provided to us by Professor Friedland concerning his work with Allan.

In Section 6, we highlight a key insight of Allan's work: that the geometrical picture of mode conversion greatly simplifies when it is viewed in ray phase space, especially in spatial dimensions higher than one. Friedland and Kaufman showed that the mode conversion transmission coefficient involves the Poisson bracket of the two uncoupled dispersion functions. The appearance of the Poisson bracket was an important clue that Allan pursued in later publications. This led to a beautiful synthesis of ideas: 1] Friedland's congruent reduction method and a local linearization of the dispersion matrix can be used to develop a local 2×2 PDE governing the two resonant modes. The approach is completely general and can be applied to any N-component wave equations, whether of integral or differential, or mixed integral/differential form. 2] Given the reduction to the local 2×2 form, one can then use various transformations to find a complete solution. These transformations are generalizations of Fourier transformations (*metaplectic* transformations) which arise in the theory of linear canonical transformations. Metaplectic transformations entered plasma theory in the work of Littlejohn concerning wave packet formulations. The use of phase space ideas provides a very natural geometric picture of what is going on and points toward generalizations. The exchange of ideas among the Berkeley plasma theory group, led by Allan and Robert Littlejohn has been very fruitful over the years. For example, Littlejohn's work on wave packet propagation, the role of the Heisenberg-Weyl group (the group of phase space shifts) and the notion of metaplectic transformations, has had a very profound influence on our approach to mode conversion.

The idea that phase space is the natural place to view conversion has informed all of Allan's subsequent work in the area, and forms the basis for our collaborative efforts with him. In Section 7, we describe some of these further efforts to develop the phase space theory of conversion by summarizing a series of papers which Allan's group of collaborators has produced, as well as other researchers in this area. These include 1] the incorporation of kinetic effects and Budden-type resonant absorption, 2] the development by our group, joined by André Jaun, of a practical algorithm for incorporating mode conversion into ray-tracing codes in realistic geometries and plasma models, and 3] a new approach to the theory of operator symbols that provides interesting perspectives on phase space representations of wave problems.

We end with a summary of active lines of research currently being pursued by our group and others.

2. Contributions to plasma wave theory prior to 1985

An INSPEC search (which only goes back to 1969) using Allan's name returned 107 hits. In this section we will give a very brief summary of the most significant papers prior to about 1985. The papers after 1985 will considered in later sections.

2.1. Manley-Rowe relations

Bruce Cohen, Ken Watson and Allan collaborated on a PRL highlighting the importance of the Manley-Rowe relations in beat heating of plasmas [1]. The following scenario was considered: two laser beams with a difference frequency that is nearly equal to the plasma frequency can resonantly interact to produce longitudinal plasma oscillations. These can then induce transitions to other transverse modes. It is the nonlinear damping of the longitudinal mode that heats the plasma. The importance of action conservation, as summarized by the Manley-Rowe relations, was emphasized. Allan returned to the subject of the Manley-Rowe relations in a later paper with Alain Brizard [2], where non-eikonal wave fields are treated. In the later paper, the Manley-Rowe relations are shown to follow from a Noether symmetry for an appropriate action principle. The use of action principles and symmetries to derive conservation laws is a recurring theme is Allan's work.

2.2. The introduction of action-angle variables for tokamak linear susceptibility and quasilinear diffusion

In a 1972 paper appearing in Physics of Fluids, Allan introduced an action-angle formalism for the study of the plasma susceptibility and quasilinear diffusion in tokamak geometry [3]. In a toroidal plasma with axial symmetry, the three adiabatically invariant actions of a particle are the magnetic moment, the canonical angular momentum, and the toroidal flux enclosed by the drift surface. Resonant interactions between particles and the normal modes of collective oscillations produce mode growth or decay and random changes in the actions. This random walk is represented by a diffusion equation in action space. Both the diffusion tensor and the growth rate depend upon a coupling coefficient which represents the work done by a normal-mode field eigenfunction on the current density of an unperturbed particle orbit. In this paper, Allan was unable to include the reaction of the background field, due to resonant diffusion, and was always concerned about this. In later work with Huanchun Ye [4] (see also Huanchun's PhD thesis) these self-consistent effects were included. By that time the use of variational principles, guiding center and oscillation center Lagrangian Lie-transforms and Noether methods had become part of Allan's toolkit.

2.3. *Hamiltonian chaos and stochastic heating*

In 1975 Gary Smith and Allan wrote a PRL on the chaotic acceleration of particles in a coherent wave field [5]. Surface-of-section plots were used to show that varying the wave amplitude leads to a transition from adiabatic to chaotic particle motion and to the onset of heating. As an application of these concepts, the authors showed the possibility of heating the tail of the ion distribution in a magnetized plasma by an intermediate-frequency acoustic wave. This paper, along with those of Charles Karney of Princeton, helped to introduce the concept of deterministic chaos into plasma theory.

2.4. *Development of the Hamiltonian Lie transform*

The development of the Hamiltonian Lie transform was carried out with John Cary (1977) and led to a 1981 paper applying this new theoretical tool to the study of ponderomotive effects [6]. This method involves the application of the Lie-transform perturbation technique to the Hamiltonian formulation of the Vlasov equation. A canonical change of coordinates is carried out to new variables in which the high-frequency has been removed. In this system the distribution function evolves according to a ponderomotive Hamiltonian, which is the kinetic generalization of the ponderomotive potential. It was shown that the ponderomotive Hamiltonian can easily be determined from the well-known linear susceptibility (a result now known as the *K-χ Theorem*).

2.5. *Beat Hamiltonians and ponderomotive potentials*

With Shayne Johnston and George Johnston, Allan recognized that the ponderomotive potential is the oscillation-center Hamiltonian [7] and a novel approach to the theory of nonlinear mode coupling in hot magnetized plasma was developed. The formulation retains the conceptual simplicity of the familiar ponderomotive-scalar-potential method, but removes the approximations. The essence of the approach is a canonical transformation of the single-particle Hamiltonian, designed to eliminate those interaction terms which are linear in the fields. The new entity (the 'oscillation centre') then has no first-order jittering motion, and generalized ponderomotive forces appear as nonlinear terms in the transformed Hamiltonian. This viewpoint is applied to derive a compact symmetric formula for the general three-wave coupling coefficient in hot uniform magnetized plasma, and to extend the conventional ponderomotive-scalar-potential method to a case with momentum dependence in the ponderomotive hamlitonian..

2.6. *The dissipative bracket*

In a 1984 paper [8], Allan showed that the concept of a Hamiltonian system (i.e., the combination of a Hamiltonian function and a Poisson-bracket structure) could be generalized to include a wide class of dissipative processes. This was an idea that was simultaneously and independently developed by Morrison [9] and Grmela [10] applied to different areas. Evolution of any observable is generated jointly by a Hamiltonian, with an entropy-conserving Poisson bracket and an entropy, with an energy-conserving dissipative bracket. This approach yields many of the standard kinetic equations such as those representing particle collisions, three-wave interactions, and wave-packet resonances.

2.7. Recognition of the fundamental importance of the action principle

Through a series of papers with several authors [11, 12, 13, 14], Allan became a firm believer in the fundamental importance of action principles. Whitham's averaged-Lagrangian method was a strong influence [15]. These ideas were developed further in Bruce Boghosian's thesis (1984-6) [16]. Action principles play a central role in these papers and provide a very elegant method for developing covariant formulations of plasma wave problems, including relativistic effects. Noether symmetries play an important role in the study of conservation laws. In addition, through use of the Lagrangian Lie transform, the action principle becomes a powerful tool for perturbation calculations.

During these years, Allan recognizes that his greatest influence was that of Robert Littlejohn, who introduced him to symplectic geometry, noncanonical brackets, Lagrangian variational principles, and the phase-space Lagrangian Lie-transform perturbation method. As we'll discuss in later sections, Robert continues to have a great influence on Allan's work.

This completes our summary of the most significant of Allan's pre-1985 papers in plasma wave theory. We now move on to the years when Allan became deeply involved in the application of WKB methods in plasma wave theory, and various extensions of WKB theory. Prior to discussing Allan's contributions, we start with a brief summary of the underlying ideas.

3. WKB theory: a very short synopsis

In these informal notes we will use the terms 'WKB theory', 'ray tracing', and 'short-wavelength asymptotics' interchangeably. The more general term 'semi-classical analysis' we will use to include extensions of WKB theory that are required in the vicinity of caustics or mode conversion, where the WKB approximation is invalid.

3.1. Preliminaries: uniform and static media

We begin by discussing the uniform case, which can be dealt with using Fourier methods. Suppose we start with a scalar wave equation of the form:

$$D(i\partial_t, -i\nabla)\psi(x;t) = 0, \tag{1}$$

where ∇ is the gradient operator and $x = (x_1, \dots x_n)$ (n, the number of spatial dimensions, will be 1, 2 or 3 depending upon the problem of interest). Note that this PDE might be of infinite order (*i.e.* a *pseudodifferential* equation). We assume that the wave operator is self-adjoint, implying that (1) can be derived using the action principle:

$$\mathcal{I} = \int dt\, d^n x\; \psi^*(x;t) D(i\partial_t, -i\nabla)\psi(x;t). \tag{2}$$

The wave equation (1) is recovered via the stationarity condition $\delta\mathcal{I}/\delta\psi^* = 0$. The use of action principles in WKB problems is a central theme in much of Allan's work as we'll see.

The wave equation is supplemented by some initial/boundary data which we will leave unspecified here, but note that satisfaction of the initial/boundary conditions requires a superposition of the plane wave solutions. Inserting the ansatz $\psi(x,t) = \exp(ik\cdot x - i\omega t)$ with $k = (k_1, \dots, k_n)$ into the wave equation (1), we arrive at the condition $D(\omega, k) = 0$, which defines the *dispersion manifold.* Because the operator $D(i\partial_t, -i\nabla)$ is self-adjoint, the function $D(\omega, k)$ will be a real function of real ω and k. We concern ourselves only with real roots of $D(\omega, k) = 0$ here, but note that complex roots are often of interest in tunneling problems.

The condition $D(\omega, k) = 0$ implies that, in the $n + 1$-dimensional space (ω, k), only those points which lie on the n-dimensional surface defined by $D(\omega, k) = 0$ can be associated with freely propagating waves. The n-dimensional dispersion manifold is thus *implicitly* defined. We can, if we choose, use local coordinates on this surface and write an *explicit* relation between ω

and $k = (k_1, \ldots, k_n)$. For example, if we choose to use $(k_1, \ldots, k_n)$ as the local coordinates on the dispersion surface, we then solve $D(\omega, k) = 0$ for ω which locally defines $\omega = \Omega(k)$, which is an example of a *dispersion relation*. The implicit condition $D(\omega, k) = 0$ can have multiple branches (even an infinite number in some cases), while a dispersion relation should be understood to refer to a particular branch (using a particular choice of local coordinates).

In uniform media, the phase level sets, $\theta(x,t) = k \cdot x - \omega t$, change with time in a simple way. By considering how level sets of the phase move with time, it is possible to identify a characteristic 'velocity' associated with their evolution. The convention is to ask how points on level sets move in a direction normal to the phase fronts themselves. The phase velocity will be denoted v_p and is defined by

$$v_p \equiv \frac{\omega}{|k|}\hat{k}. \tag{3}$$

The phase velocity will play an important role in our discussion of mode conversion because resonant interaction between two wave types requires phase matching, hence they must have the same phase velocities.

Now consider a nearly monochromatic wavetrain, or a wave packet, with a well defined carrier oscillation, but still in a uniform and static background. This type of solution can be constructed by superposing plane wave solutions that lie close to one another on the same branch of $D(\omega, k) = 0$. Denote the carrier as (ω_0, k_0) (noting that $D(\omega_0, k_0) = 0$), and assume a new ansatz of the form

$$\psi(x,t) = A(x,t)e^{ik_0 \cdot x - \omega_0 t}. \tag{4}$$

Inserting this into the wave equation (1) we find that the amplitude $A(x,t)$ must satisfy

$$D(\omega_0 + i\partial_t, k_0 - i\nabla)A(x,t) = 0. \tag{5}$$

This is still completely general, we have simply carried out a change of the dependent variable from $\psi \to A$. If we now insist that $A(x,t)$ is a slowly varying function of its arguments, then we can expand the wave operator in powers of the derivatives to find at leading order:

$$D(\omega_0 + i\partial_t, k_0 - i\nabla)A(x,t) \approx \left[\frac{\partial D}{\partial \omega}\partial_t - \frac{\partial D}{\partial k} \cdot \nabla\right] A(x,t) \approx 0, \tag{6}$$

where we have used $D(\omega_0, k_0) = 0$ and the derivatives of $D(\omega, k)$ are evaluated at (ω_0, k_0). Note that $\partial D/\partial k = (\partial D/\partial k_1, \ldots, \partial D/\partial k_n)$. Provided $\partial D/\partial \omega \neq 0$, we can recast the evolution equation for the envelope into a standard advection equation:

$$[\partial_t + v_g \cdot \nabla] A(x,t) = 0, \qquad v_g \equiv -\left(\frac{\partial D}{\partial \omega}\right)^{-1} \frac{\partial D}{\partial k}, \tag{7}$$

where the *group velocity*, denoted v_g, has appeared. Note that the group velocity is associated with the evolution of the amplitude (equivalently the envelope) of the wave, hence it determines how energy and action are transported by the wave. Note, also, the narrow-banded assumption is critical. For dispersive systems, broad banded solutions do not have well-defined signal pathways.

The generalization of the above discussion to N-component wave equations proceeds as follows: The wave equation is now of the form

$$\mathsf{D}(i\partial_t, -i\nabla) \cdot \Psi(x,t) = 0, \tag{8}$$

or, in component form:

$$\sum_{n=1}^{N} \mathsf{D}_{mn}(i\partial_t, -i\nabla)\psi_m(x,t) = 0. \tag{9}$$

We still assume self-adjointness. This system has plane-wave solutions of the form: $\hat{\mathbf{e}}(\omega, k) \exp(ik \cdot x - i\omega t)$. Inserting this ansatz, the wave equation becomes:

$$\mathsf{D}(\omega, k) \cdot \hat{\mathbf{e}}(\omega, k) = 0, \tag{10}$$

where $\mathsf{D}(\omega, k)$ is a self-adjoint $N \times N$ matrix for all real ω and k. Therefore, plane-wave solutions of the wave equation can only exist for those values of (ω, k) for which $\mathsf{D}(\omega, k)$ has a null eigenvalue ($\hat{\mathbf{e}}(\omega, k)$ is the associated eigenvector). We can test for a null eigenvalue by taking the determinant, leading to the definition of the dispersion manifold for multicomponent wave equations:

$$D(\omega, k) \equiv \det(\mathsf{D}(\omega, k)) = 0, \tag{11}$$

a result that is both familiar and deceptive because by moving so quickly we have slipped some things under the rug. Let's proceed a bit more slowly.

First, the dispersion matrix $\mathsf{D}(\omega, k)$ is well-defined at each (ω, k) and is self-adjoint everywhere, with well-defined eigenvalues and eigenvectors. Therefore, at each (ω, k) we can decompose the dispersion matrix onto an orthonormal basis composed of its N eigenvectors (*i.e.* solutions of $\mathsf{D} \cdot \hat{\mathbf{e}}_\alpha = \lambda_\alpha \hat{\mathbf{e}}_\alpha$):

$$\mathsf{D}(\omega, k) = \sum_{\alpha=1}^{N} \lambda_\alpha(\omega, k) \hat{\mathbf{e}}_\alpha(\omega, k) \hat{\mathbf{e}}_\alpha^\dagger(\omega, k). \tag{12}$$

Generically, the eigenvalues of a hermitian matrix will be non-degenerate implying the N eigenvectors are automatically orthogonal. If two (or more) of the eigenvalues of $\mathsf{D}(\omega, k)$ are degenerate, however, things get more interesting. This is because if $\hat{\mathbf{e}}_1$ and $\hat{\mathbf{e}}_2$ have the same eigenvalue, then so does any linear combination of them. Therefore, it would appear that we have complete freedom in choosing a basis in the degenerate subspace. Mathematically speaking, this is certainly true for any fixed hermitian matrix. However, we must always keep in mind that we are dealing with a family of hermitian matrices, namely one for each point (ω, k). If the matrix is degenerate at the point (ω, k), then it generically will *not* be degenerate at a neighboring point $(\omega + \Delta\omega, k + \Delta k)$. At this neighboring point there is therefore *no* freedom to choose the eigenvectors for decomposition. (More precisely, the only freedom we have is the choice of phase for $\hat{\mathbf{e}}(\omega, k)$.) Hence, at a point of degeneracy it would seem reasonable that one should choose eigenvectors that smoothly connect to their non-degenerate neighbors. (This is the essential idea behind degenerate perturbation theory in quantum mechanics.) In one-dimensional parameter spaces this approach leads to a well-defined choice of basis vector at the point of degeneracy. But in higher dimensions it does not because the result often depends sensitively upon the direction one uses to approach the degeneracy. Hence, there is no choice of polarization eigenvector at the point of degeneracy that will smoothly connect to all of the surrounding non-degenerate neighbors.

In nonuniform media, this nasty behavior near a degeneracy is precisely what leads to the breakdown of the WKB approximation. This is because, as we discuss in the following section, in vector WKB the polarization is assumed to be transported adiabatically following a ray. This polarization vector is usually taken to be a local null-eigenvector of the dispersion matrix, following the intuitively reasonable argument that what is true globally for the uniform case should be true locally in the nonuniform one. However, in the vicinity of a degeneracy, where two or more of the eigenvalues of the dispersion matrix approach zero, the null eigenvectors are badly behaved for precisely the same reason as in the uniform case. Therefore, they change *rapidly* following a ray and the adiabatic assumption breaks down. We will discuss at some length methods that have been developed to overcome this difficulty while retaining the power of WKB methods away from the degeneracy.

3.2. Non-uniform and time-varying media

All of the material in this section has been covered in great detail in earlier papers [19, 18, 17]. We include a brief heuristic introduction to the ideas for completeness.

Suppose the background plasma is no longer uniform in space and that it varies in time as well. This significantly complicates the problem. However, if we can treat these new effects as a perturbation, then we can use plane waves as local solutions with slowly varying parameters. Intuitively, what we would like to do is argue that the dispersion function $D(\omega, k)$ now becomes local, $D(t, \omega, x, k)$. The correspondence $k \to -i\nabla$ still holds. The new complication is that ∇ does not commute with x so we must be very careful about ordering issues. The Weyl symbol calculus provides a systematic approach for dealing with these matters, and we'll discuss it momentarily. Let's write the scalar wave equation (1) in the following form:

$$D(t, i\epsilon\partial_t, x, -i\epsilon\nabla)\psi(x,t) = 0. \tag{13}$$

The formal expansion parameter ϵ has been introduced to clarify the asymptotic balances about to be invoked. The Weyl calculus ensures that this wave operator is symmetrized (for example the product xk corresponds to the operator $(\hat{x}\hat{k} + \hat{k}\hat{x})/2$). We still assume the wave operator is self-adjoint, therefore there is an action principle:

$$\mathcal{I} = \int dt\, d^n x\, \psi^*(x,t) D(t, i\epsilon\partial_t, x, -i\epsilon\nabla)\psi(x,t) = 0. \tag{14}$$

The wave equation (13) arises from the stationarity condition $\delta\mathcal{I}/\delta\psi^* = 0$.

Guided by the earlier calculation for the uniform medium, we insert the ansatz $\psi(x,t) = A(x,t)\exp(i\theta(x,t)/\epsilon)$ and find:

$$D(t, -\theta_t + i\epsilon\partial_t, x, \nabla\theta - i\epsilon\nabla)A(x,t) = 0. \tag{15}$$

At this stage no approximations have been made, we have only changed the dependent variable. Expand in powers of ϵ to find:

$$\left[D(t, -\theta_t, x, \nabla\theta) + \epsilon\frac{i}{2}\left(\frac{\partial D}{\partial\omega}\partial_t + \partial_t\frac{\partial D}{\partial\omega}\right) - \epsilon\frac{i}{2}\left(\frac{\partial D}{\partial k}\cdot\nabla + \nabla\cdot\frac{\partial D}{\partial k}\right)\right]A(x,t) = \mathcal{O}(\epsilon^2). \tag{16}$$

We now insist that these expressions vanish order-by-order in ϵ.

At $\mathcal{O}(\epsilon^0)$ we have $D(t, -\theta_t, x, \nabla\theta) = 0$, which is a nonlinear PDE (the *Hamilton-Jacobi equation*) for the unknown phase function $\theta(x,t)$. Defining the local frequency $\omega \equiv -\theta_t$ and wavevector $k = \nabla\theta$ we see that we can interpret the Hamilton-Jacobi equation as a requirement that the local phase function must be chosen so the local dispersion relation $D(t, \omega, x, k) = 0$ is satisfied.

At $\mathcal{O}(\epsilon)$ we get a linear PDE for the envelope $A(x,t)$. It should be emphasized, however, that the coefficients of this PDE involve the derivatives of the dispersion function at the points $\omega = -\theta_t$ and $k = \nabla\theta$, and hence it appears to require that the solution of the Hamilton-Jacobi equation is already known. In fact, the situation isn't quite so bad. This is because both the Hamilton-Jacobi equation and the amplitude transport equation can be solved using ray tracing methods (following directly in Hamilton's footsteps). Once the *rays* are known, the equation for the envelope can be recast as an ODE describing how the amplitude changes along any given ray.

It must be emphasized that in spatial dimensions higher than one it is not sufficient to follow only single rays. One must follow an entire ray family in order to transport phase and amplitude. We will discuss this important point further when we discuss modern developments because there is a very natural geometrical interpretation of this notion that leads to the concept of a *Lagrange manifold.*

The first order equation (16) can be recast as a conservation law for wave action by brute force. However, Allan showed that it is possible to derive this in a more elegant fashion, and to gain important insight into the origin of the conservation law by returning to the action principle (14). Insert the eikonal ansatz directly into the action principle and keep only the leading order terms. This leads to the modified action principle:

$$\mathcal{I}' = \int dt\, d^n x\, D(t, -\theta_t, x, \nabla\theta)|A|^2(x,t) = 0. \tag{17}$$

Stationarity under the variation $\delta\mathcal{I}/\delta A^* = 0$ gives the Hamilton-Jacobi equation. The action principle is invariant under a global shift by a constant phase $\theta \to \theta + \theta_0$. Therefore, by Noether's theorem, there is an associated conserved quantity. This conserved quantity is the wave action, and the associate wave action conservation law, which is usually derived by direct calculation, is revealed to be due to an underlying symmetry of the theory.

Returning for the moment to the rays: the rays equations are written in terms of a ray parameter σ (for a derivation, see the papers cited at the beginning of this section):

$$\dot{x} \equiv \frac{dx}{d\sigma} = -\frac{\partial D}{\partial k}, \qquad \dot{k} \equiv \frac{dk}{d\sigma} = \frac{\partial D}{\partial x}. \tag{18}$$

Recall that x and k are vectors. The ray parameter and the physical time are related via $dt/d\sigma = \partial D/\partial\omega$. The phase increment $\Delta\theta$ following a ray that starts at the point x_0 with wavevector k_0 is given by first following the evolution generated by Hamilton's equations to find $(x(\sigma; x_0, k_0), k(\sigma; x_0, k_0))$. This information is now used to compute $\dot{x}$ along the ray and an integration performed to find $\Delta\theta = \int d\sigma k \cdot \dot{x}$. Because Hamilton's equations satisfy uniqueness (at least they do for the smooth types of dispersion functions of most interest in physics), rays will not cross in phase space. However, they can cross in x-space. This leads to singular behavior in the amplitude transport equation (caustics). Therefore, in general, the Hamilton-Jacobi equation does not have global solutions, only local ones. The treatment of caustics is discussed in [20].

WKB theory for vector wave equations can be brutally summarized by saying that if we now start with a wave equation of the form:

$$\mathsf{D}(t, i\epsilon\partial_t, x, -i\epsilon\nabla) \cdot \Psi(x;t) = 0; \tag{19}$$

we insert an ansatz of the form:

$$\Psi(x;t) = e^{i\theta(x;t)/\epsilon} A(x,t)\hat{\mathbf{e}}(x;t). \tag{20}$$

At leading order in ϵ this requires:

$$\mathsf{D}(t, \omega = -\theta_t, x, k = \nabla\theta) \cdot \hat{\mathbf{e}}(x;t) = 0. \tag{21}$$

This requires that $\theta(x;t)$ is such that $\mathsf{D}(t, \omega = -\theta_t, x, k = \nabla\theta)$ has a null eigenvalue, with $\hat{\mathbf{e}}(x;t)$ the associated null eigenvector. The existence of a null eigenvalue can be tested for using the determinant, hence the Hamilton-Jacobi equation for multicomponent wave problems is:

$$D(t, \omega = -\theta_t, x, k = \nabla\theta) \equiv \det(\mathsf{D}(t, \omega = -\theta_t, x, k = \nabla\theta)) = 0. \tag{22}$$

As in the scalar case, this equation in general has only local solutions, so something must be done to connect the local solutions together to find the global solution. We note also that the condition (21) determines the direction of $\hat{\mathbf{e}}(x,t)$, but not its phase. A very elegant approach to this topic is one of Allan's many contributions that we discuss in later sections.

Before ending this very rudimentary discussion, we note that the ideas presented here are also readily applicable in the quantum setting and what we have described is essentially the approach taken in adiabatic perturbation theory. These methods underly what has come to be called EBK quantization, but it must be supplemented by methods for dealing with the localized breakdown of the WKB approximation. These breakdowns occur in two different ways:

(i) Near caustics the family of rays used to construct the solutions have bad projections from phase space back onto configuration space, hence the phase assignment at x is not well-defined and the amplitude can become be singular. A variety of authors have contributed to the theory here, and we point the interested reader to [20] as a point of entry to the literature.

(ii) Near degeneracies, the null eigenvectors of $\mathsf{D}(t,\omega,x,k)$ will not be well-behaved and will change significantly with slight changes of position in phase space (see the earlier discussion for the uniform medium vector theory). In the quantum mechanics literature, this type of breakdown is often called a 'non-adiabatic' effect. The literature there stretches back to the 1930's in the work of Landau, Zener, Stückelberg and Teller. Reviews of this older literature have appeared recently [21, 22] and Littlejohn has recently discussed Teller's contributions to this topic [23]. In the plasma literature this phenomena is known as *mode conversion*. We emphasize that the underlying mathematics of these problems are essentially the same, even though the physical interpretations are very different. We will discuss mode conversion much more in a later section because it has been a major focus of Allan's work in the last twenty years.

Up to this point we have been pursuing a very heuristic approach to the topic in order to emphasize that the ideas are very intuitive and flow directly from experience with the theory of uniform media. The time has come to discuss some of the more modern ideas that have been introduced. Allan's work in plasma wave theory for the last twenty years has been informed by these newer ideas, which put the intuitive approach onto a much firmer mathematical foundation.

3.3. Semi-classical extensions: Geometrical ideas come to the fore

The field of semi-classical analysis has developed significantly since the 1960's and the literature is wide and varied. It is very much alive as a field of research in mathematics and theoretical physics. Some of the ideas reach back to the 1930's in the work of Moyal, Weyl and the later work of Wigner. But, some of the ideas are newer, especially the geometric flavor of the modern approach. In fact, semi-classical analysis is an area where the mathematical fields of geometry, algebra and analysis come together. What follows is an admittedly skewed picture of this field, colored by the authors' particular viewpoint. No attempt has been made to carry out a full survey, which would have resulted in a very different paper, and we apologize to those whose work is not mentioned. Here we simply summarize some of the key ideas that will be of use describing Allan's work in the area.

3.3.1. Weyl symbols of operators In his book on group theoretical methods in quantum mechanics [24], Weyl described a method for associating operators on Hilbert space with functions on classical phase space:

$$\hat{A} \leftrightarrow a(q,p). \tag{23}$$

Recall that a linear operator is defined abstractly by its action on vectors that live in a Hilbert space. It can be *represented* by a matrix if we choose a particular basis in the space. The symbol is simply a different way of representing the operator, but now on the vector space of functions on classical phase space.

Journal of Physics: Conference Series **169** (2009) 012008 doi:10.1088/1742-6596/169/1/012008

The mapping $\hat{\mathcal{A}} \leftrightarrow a(q,p)$ can be given explicitly. For example, using the q-representation of the operator we can express the symbol as:

$$a(q,p) = \int d^n s \, e^{-ip\cdot s} \langle q + \frac{s}{2} | \hat{\mathcal{A}} | q - \frac{s}{2} \rangle. \tag{24}$$

In this formula, and in what follows, we ignore normalizing factors of 2π in order to focus on the conceptual relations and use the notation $q = (q_1, \dots q_n)$ and $p = (p_1, \dots p_n)$. Depending upon the circumstance, we might also use an extended phase space to include time and frequency.

To go backward from the symbol to the operator, we first take the Fourier transform of the symbol:

$$\tilde{a}(\sigma,\tau) = \int d^n q d^n p \, e^{i(\sigma\cdot q + \tau\cdot p)} a(q,p). \tag{25}$$

Now use the transform of the symbol as a set of expansion coefficients over the set of phase space shifts generated by the operators $\hat{q}$ and $\hat{p}$. In essence, we do an inverse Fourier transform but, instead of using q and p, use the operators $\hat{q}$ and $\hat{p}$:

$$\hat{\mathcal{A}} = \int d^n \sigma d^n \tau e^{-i(\sigma\cdot\hat{q} + \tau\cdot\hat{p})} \tilde{a}(\sigma,\tau). \tag{26}$$

Technicalities aside, the mapping $\hat{\mathcal{A}} \leftrightarrow a(q,p)$ is essentially one-to-one and invertible. This is extremely useful, because it allows us to gain insight into the action of the operator $\hat{A}$ by considering its 'classical' counterpart $a(q,p)$. The mapping is also 'topological' in the sense that it preserves the notion of closeness: operators that are 'close' will map to symbols that are 'close'. Hence, we can develop approximations to operators by using series expansions in the related space of symbols.

Because this mapping preserves neighborhood relationships, it must also reflect the fact that operators generally don't commute. Hence, the symbol of the operator $\hat{\mathcal{A}}\hat{\mathcal{B}}$ cannot be equal to the symbol of $\hat{\mathcal{B}}\hat{\mathcal{A}}$. It is possible, however, to derive by direct calculation using the definition (24) how the symbol of the product $\hat{\mathcal{A}}\hat{\mathcal{B}}$ can be written in terms of the symbols of $\hat{\mathcal{A}}$ and $\hat{\mathcal{B}}$, denoted $a(q,p)$ and $b(q,p)$ respectively, a result that goes back to Moyal. See, for example, Steve McDonald's Physics Report on the subject [25]. The result is:

$$\hat{\mathcal{A}}\hat{\mathcal{B}} \leftrightarrow a(q,p) * b(q,p), \tag{27}$$

where the '$*$' product can be written as a Moyal series:

$$a(q,p) * b(q,p) \equiv a(q,p) \; \exp\left[\frac{i}{2}\left(\overleftarrow{\partial_q}\cdot\overrightarrow{\partial_p} - \overleftarrow{\partial_p}\cdot\overrightarrow{\partial_q}\right)\right] b(q,p). \tag{28}$$

This result is completely general and relies only upon the formal definition of the mapping from operators to symbols. As we will see in Section 4.2, it was used in a very elegant way by McDonald and Kaufman to derive the wave kinetic equation for incoherent electromagnetic fields including weak sources and sinks [26]. To our knowledge, this was the first application of Weyl's ideas in plasma wave theory.

We note a very important example of an operator symbol. If we have a wave function $|\Psi\rangle$ in Hilbert space and construct its projector $|\Psi\rangle\langle\Psi|$, then using the definition (24) a short calculation shows that the symbol of this operator is the *Wigner function*. This function has properties that make it useful for the study of waves in phase space. This will be covered in a moment.

Before moving on, we mention that our colleague Nahum Zobin has developed a theory of symbols based upon group theoretical ideas which shows that many results in the symbol calculus have a group theoretic origin. The product rule for symbols, for example, is simply a convolution theorem, but for a non-commutative group: the Heisenberg-Weyl group. (The more familiar convolution theorem from Fourier analysis is related to the translation group on configuration space, which is commutative.) The interested reader is referred to Steve Richardson's PhD. thesis [27].

3.3.2. Pseudodifferential operators The Weyl symbol calculus provides a further useful tool for the study of plasma waves: it allows us to convert integral operators to pseudodifferential form. In this form, the application of WKB methods becomes far more transparent. Suppose we start with a vector wave equation written in integral form:

$$\sum_m \int dt' d^n x' \mathsf{D}_{nm}(t,t',x,x')\psi_m(x',t') = 0. \tag{29}$$

Is is not immediately apparent how to apply WKB methods to this system. However, by first computing the symbol of the kernel using the Weyl prescription (here we have to use the extended phase space to include $(q,p) = (t,x,\omega,k)$), it is possible to construct a pseudifferential operator whose action on the Hilbert space of wave functions $|\Psi\rangle$ is essentially the same as that of the original integral form.

A simple example of this idea can be given. The shift operator, whose action on a wavefunction $\psi(x)$ is defined via $\left[\hat{T}_a\psi\right](x) \equiv \psi(x-a)$ can also be expressed in pseudodifferential form:

$$\psi(x-a) = e^{-a\partial_x}\psi(x), \tag{30}$$

where the right hand side (the Taylor series) is written in pseudodifferential form.

Using the Weyl calculus, the integral form of the wave equation can be recast into the pseudodifferential form:

$$\sum_m \mathsf{D}_{nm}(t, i\partial_t, x, -i\nabla)\psi_m(x,t) = 0, \tag{31}$$

as assumed in earlier sections. This can be understood as simply a different way of characterizing the same operator. See [28] for details. To our knowledge, Allan Kaufman and Robert Littlejohn and their students were the first to bring these ideas into plasma physics.

We mentioned earlier the important fact that the symbol-operator mapping preserves notions of closeness. If it did not it would be of limited use. This suggests the following question: suppose we Taylor expand the symbol of $\hat{\mathcal{A}}$ about a fixed, but arbitrary, point in phase space:

$$a(q,p) \approx a(q_0,p_0) + (q-q_0)\cdot a_{q0} + (p-p_0)\cdot a_{p0} + \ldots \tag{32}$$

Now truncate this expansion. This is a new symbol and it has a related operator, $\hat{\mathcal{A}}'$, that is close to the original operator in some sense:

$$\hat{\mathcal{A}}' = a(q_0,p_0)\,\mathsf{Id} + (\hat{q}-q_0)\cdot a_{q0} + (\hat{p}-p_0)\cdot a_{p0}, \tag{33}$$

where Id is the identity operator in the relevant Hilbert space. The operators $\hat{q}$ and $\hat{p}$ are multicomponent position and momentum operators. For example, in the q-representation $\hat{q}$ becomes multiplication by (t,x) and $\hat{p}$ becomes $(i\partial_t, -i\nabla)$. We now ask the question: In what sense is $\hat{\mathcal{A}}'$ close to $\hat{\mathcal{A}}$? In ordinary function approximation, truncation of Taylor series leads to another function that is 'close' to the original one only within some region. What is the related region where $\hat{\mathcal{A}}$ and $\hat{\mathcal{A}}'$ agree? The region of agreement lies in the Hilbert space where the two operators act. The two operators 'agree' in their action on wave functions $|\psi\rangle$ that lie in that region. These functions are local plane waves, or more precisely, they are wave packets. For example, in the q-representation they are wave functions of the form:

$$\langle q|\psi\rangle = \tilde{\psi}(q-q_0)e^{ip_0\cdot(q-q_0)} \tag{34}$$

where the envelope $\tilde{\psi}$ is a smooth function of its argument. This notion can be developed much further. By an appropriate choice of envelope function it is possible to construct complete sets

Journal of Physics: Conference Series **169** (2009) 012008 doi:10.1088/1742-6596/169/1/012008

of states. Using the Weyl calculus, the action of any operator can be then decomposed into its much simpler local action (in Hilbert space) on each of these wavepackets. This makes wave evolution look much more like the evolution of a classical gas. We refer the reader to Littlejohn's 1986 review of his wavepacket formalism [28].

Thus, symbols provide a powerful means for constructing approximations to general operators, but the approximate form of the operator must be understood to act on wave functions that are localized in a particular region of phase space. In the mathematics literature, this is called *microlocal analysis* because it is local both in q and p.

3.3.3. Eikonal waves and Lagrange manifolds The previous discussion concerning symbols and pseudodifferential operators was very general. Now we consider a very special class of solutions of the wave equation: those which are of eikonal form. This introduces several important new ideas. It has been known since the time of Hamilton that in order to solve multidimensional wave equations using ray methods one must use a family of rays. The initial conditions for this family are determined by the initial/boundary conditions of the wave equation, and these conditions must be consistent with the search for a WKB solution. (For example, if we drive an antenna at the boundary of a plasma with a broadband space and time dependence then we should not expect to find the solution in the interior of the plasma to be of the narrowbanded eikonal form). The rays that evolve from WKB-type initial conditions (using Hamilton's equations), fill out a smooth surface that has a dimension that is always equal to the dimension of the space (t, x), and hence one-half the dimension of the full phase space (t, x, ω, k). This is because eikonality, which assumes there is a well-defined phase function $\theta(x, t)$, implies that at each (x, t) there are a well-defined frequency and wave-vector: $(\omega(x, t) = -\theta_t, k(x, t) = \nabla\theta)$. This is valid so long as $\theta(x, t)$ is well-defined. What happens when this is no longer true? At caustics the ray evolution in phase space is still well-defined, but a problem has developed in the projection of the ray family back down to x-space. It is reasonable to ask, since the ray family is perfectly well behaved in phase space, whether it might be possible to choose a different projection (say to k-space) to find the way around the problem. This turns out to be possible.

Arnold introduced the concept of a *Lagrange manifold* to capture this essential idea: a Lagrange manifold is a surface in phase space that has a dimension that is one-half that of the full phase space and at each point on the surface the tangent vectors all have zero symplectic product. This implies that the surface locally looks like a configuration space. The notion of a Lagrange manifold is now a central idea in the geometric theory of Hamiltonian mechanics and semiclassical analysis. Simply put, Lagrange manifolds are the only surfaces in phase space that could be associated with ray families of some wave equation. This is a new idea that is needed in multidimensional problem but is a triviality in one spatial dimension. In one spatial dimension, with a two-dimensional phase space, a ray *is* a lagrange manifold. It is also fills the dispersion surface. In spatial dimensions greater than one, however, this is no longer true. In $2n$-dimensional phase space, a ray is still one-dimensional, but a Lagrange manifold is n-dimensional and a dispersion manifold $(2n - 1)$-dimensional.

As mentioned earlier, WKB theory will break down near caustics. However, the Lagrange manifold associated with the ray family is perfectly well behaved. Even though it might develop folds or other caustic singularities under projection, it will always remain a Lagrange manifold because the surface is generated by Hamilton's equations. This allows for a very systematic and elegant approach to classifying the types of caustic singularities that can develop and local normal forms for their analysis [20].

With these ideas in place, we now return to Allan's post-1985 work in plasma wave theory. Following Lazar Friedland's suggestion, we break this into 'pre'- and 'post'-conversion.

Journal of Physics: Conference Series **169** (2009) 012008 doi:10.1088/1742-6596/169/1/012008

4. Prior to 'conversion'

4.1. Wave chaos in stadia

In a 1979 PRL Allan and Steve McDonald described solutions of the two-dimensional Helmholtz equation for a stadium-shaped cavity [29]. The underlying classical system is chaotic, and the question arose of how the related quantum system behaved. In this context, the notion of 'quantum chaos' concerns the nature of wave functions and eigenvalues when the short-wave-limit Hamiltonian has stochastic trajectories. In the 1979 PRL, it was shown that the probability distribution of eigenvalue separations, $P(\Delta E)$, is a Wigner distribution (characteristic of a random Hamiltonian), in contrast to the clustering found for a separable equation. The eigenfunctions exhibited a random pattern for the nodal curves, with isotropic distribution of local wave vectors. This was among the first papers in what became the very active area of quantum chaos. It was followed almost a decade later by a much longer paper giving more details of the calculations [30]. The local spatial correlation function $\langle \psi_n(x+\frac{1}{2}s)\psi_n(x-\frac{1}{2}s)\rangle$ and the probability distribution $P_n(\psi)$ of wave amplitude for normal modes ψ_n were computed and compared with predictions based on semiclassical arguments applied to the nonintegrable classical Hamiltonian.

4.2. Allan's work with Steve McDonald on the wave-kinetic equation

As mentioned earlier, to our knowledge Allan Kaufman and Robert Littlejohn were the first to use ideas from the Weyl calculus in plasma wave theory. Allan and Steve McDonald used these concepts in a very elegant paper [31] to show how to derive the wave-kinetic equation for incoherent electromagnetic waves. Because we have already described the Weyl product law for operators, we can describe this important paper very succinctly. Suppose we start with a wave equation for the electric field, written in abstract form as:

$$\sum_{m=1}^{N} \hat{\mathsf{D}}_{nm}|E_m\rangle = |j_n\rangle; \qquad n = 1, 2, 3, \tag{35}$$

where sources have been included on the right hand side. We use the Dirac notation because the logic of the approach is clearer. For concreteness, we can construct the x-representation of the wave equation by projecting onto $\langle x, t|$ and inserting a complete set of states:

$$\sum_{m=1}^{3} \int dt' d^n x' \langle x, t|\hat{\mathsf{D}}_{nm}|x', t'\rangle\langle x', t'|E_m\rangle = \langle x, t|j_n\rangle, \qquad n = 1, 2, 3; \tag{36}$$

or

$$\sum_{m=1}^{3} \int dt' d^n x' \mathsf{D}_{nm}(t, t', x, x') E_m(x'; t') = j_n(x; t), \qquad n = 1, 2, 3. \tag{37}$$

The wave operator has an anti-hermitian part due to dissipation, as required by the Kramers-Kronig relations. The dissipation is treated perturbatively. Returning to the abstract notation, the adjoint form of the wave equation is:

$$\sum_{m=1}^{N} \langle E_m|\hat{\mathsf{D}}^{\dagger}_{mn} = \langle j_n|; \qquad n = 1, 2, 3. \tag{38}$$

Separating the wave operator into hermitian and anti-hermitian pieces (denoted D' and D'' respectively) and introducing a formal small parameter to keep track of orderings, we have (keeping in mind that $|E\rangle$ and $|j\rangle$ are 3-component objects):

$$\left[\hat{\mathsf{D}}' + \epsilon\hat{\mathsf{D}}''\right] \cdot |E\rangle = \epsilon|j\rangle. \tag{39}$$

We can take the outer product with the adjoint equation to get, after a little manipulation:

$$\left[\hat{\mathsf{D}}' + \epsilon\hat{\mathsf{D}}''\right] \cdot |E\rangle\langle E| = \epsilon|j\rangle\langle j| \cdot \left[\hat{\mathsf{D}}' + \epsilon\hat{\mathsf{D}}''\right]^{-1}. \tag{40}$$

This is an equation for the spectral tensor of the electric field $|E\rangle$, written in a representation-free manner.

Kaufman and McDonald now use the Weyl calculus to compute the symbol of this equation. They then invoke the Moyal formula (28) for the product of symbols. This expresses the equation in terms of the spectral tensors of $E(x;t)$ and the source $j(x;t)$, now given as distributions (Wigner tensors) on phase space. Up to this point, the calculation is completely general, though formal. They now proceed to use the small parameter ϵ to set up an asymptotic expansion. All quantities are expanded in powers of ϵ and like powers collected. Weak sources and weak dissipation means the conservative part of the free evolution dominates at leading order. One still must deal with the fact that the Moyal expansion is infinite order in derivatives. If the spectral tensors are smooth functions on phase space, then one can introduce an ordering in the Moyal series which assumes each order of derivative introduces a power of ϵ. Strictly speaking, this is not true. Wigner functions are typically distributions, not smooth functions. However, for a scalar problem if the wave field is incoherent, rather than eikonal, the Wigner function can be smooth *along* the dispersion manifold, even if it is still sharply confined to the dispersion surface itself. This is because the fine structure off the dispersion manifold that appears in the Wigner function for coherent waves is assumed to phase mix away when incoherent fields are used. Therefore, the Moyal series is assumed to have a natural ordering in ϵ at each order of derivative.

The symbol of $\hat{\mathsf{D}}'$ is the dispersion matrix $\mathsf{D}'(t,\omega,x,k)$. It is self-adjoint at each point in the phase space and assumed to be a smooth function of its arguments. Therefore, at each point in phase space it has a set of real eigenvalues that are well defined. Because of the assumption that the sources are weak, at leading order it can be assumed that waves must lie on a dispersion manifold where one of the eigenvalues is zero. Kaufman and McDonald assume that there are no degeneracies (hence mode conversion is not allowed in this treatment). With these assumptions, it is possible to simplify the vector problem to a scalar one by projecting onto the null subspace associated with the mode in question. Then, assuming that the distribution function fills the dispersion manifold and is smooth along it, they show that the Moyal series has as its next order term a transport equation. They identify this as the wave-kinetic equation for the wave action.

4.3. Phase-space action principles

In 1987 Allan wrote a paper describing a phase-space Lagrangian action principle and the generalized $K - \chi$ Theorem [32]. In this paper the covariant coupled equations equations for plasma dynamics and the Maxwell field were expressed using a phase-space action principle. Carrying out a gauge-invariant Lagrangian Lie transform, the linear interaction is transformed to the bilinear beat Hamiltonian. This leads directly to the generalized linear susceptibility.

4.4. Covariant vector WKB

Many of the research ideas that Allan and co-workers had pursued in the previous decade came together in a very beautiful and exceedingly concise paper with Huanchun Ye and Yukkei Hui [33]. In this letter a Lorentz-covariant action principle for vector waves is introduced. Using the methods of the Weyl symbol calculus, this action principle is then rewritten in terms of quantities defined on ray phase space (*e.g.*, the Wigner tensor and the dispersion tensor). Insertion of the eikonal ansatz is then shown to lead to the hamilton ray equations which define the Lagrange manifold of associated rays. At next order the wave action conservation law is recovered and shown to be due to a Noether symmetry. In addition, a Lorentz-covariant

Journal of Physics: Conference Series **169** (2009) 012008 doi:10.1088/1742-6596/169/1/012008

transport law for the polarization phase is derived. The polarization phase transport has an interesting mathematical structure, and these results were followed up by Littlejohn and Flynn who showed that, although the phase transport equations themselves were not gauge invariant, the resulting quantization condtions were [34].

In the work described so far resonances were excluded from consideration. Removing this limitation leads us to the next topic: mode conversion.

5. The conversion to mode conversion

When two or more of the eigenvalues of the dispersion matrix are nearly zero the associated null eigenvectors will depend sensitively upon their position in the ray phase space, for reasons described in previous sections. The WKB approximation breaks down because of this and a new type of asymptotic approximation must be developed near the degeneracy. It is assumed that WKB is valid away from the region of degeneracy, hence the problem reduces to 1] finding the appropriate local description of the wave dynamics in the vicinity of the conversion, and 2] fitting this local solution onto the incoming and outgoing WKB solutions. The picture is that an incoming ray 'splits' into two outgoing rays and some type of connection formula must be applied ray-by-ray. In the immediate vicinity of the conversion, however, the local wave fields will not be well-described by the WKB solutions, so this simple picture should be viewed as heuristic only.

There is, of course, a huge literature on this topic because it occurs throughout physics. It can be found to play a role in fields as diverse as RF heating of fusion plasmas [35], ionospheric physics [36], ocean waves [37, 38, 39], magnetohelioseismology [40], atomic, molecular and optical physics [41], and neutrino physics [42]. While there is a large physics literature on conversion in one dimension (see [43, 44, 45, 46, 47, 48] and references therein), and WKB-type methods were applied to multicomponent wave equations as early as Rayleigh, there has been comparatively little attention given to multidimensional conversion in multicomponent wave equations. Exceptions to this in the physics literature are the papers of Bernstein and Friedland [49], Friedland [50], Friedland and Kaufman [51], Littlejohn and co-workers [52], Tracy and Kaufman [53, 54], and Krasniak and Tracy [55, 56], and Nassiri-Mofakham [57]. In the mathematics literature, of particular note is the work of Braam and Duistermaat [58, 59], Colin de Verdiére [60, 61], and Emmrich and Weinstein [62, 63].

In the following sections we will recount Allan's 'conversion to conversion', as recalled by Lazar Friedland who played a key role in exciting Allan's interest in the topic, followed by a discussion of some of the applications and further developments that followed Allan's conversion.

Allan's conversion by Lazar Friedland

This section is adapted directly from notes provided by Lazar Friedland in the Fall of 2007. Lazar writes:

> "In 1979-1982 I joined Ira Bernstein at Yale to work on further development of his general geometric optics (multi-dimensional WKB) formalism in space/time varying plasmas [49]. Formally, the geometric optics theory comprised a perturbation expansion with respect to the small parameter characterizing the slow variation of the background plasma. The perturbation scheme used a formal decomposition of the local (Hermitian) dielectric tensor in the wave propagation problem in terms of its eigenvectors and eigenvalues. A particular geometric optics mode was associated with vanishing of one of the eigenvalues (the nondegenerate case) and both the amplitude and the eikonal phase of the wave could be calculated by solving a set of ordinary differential equations along the characteristics (geometric optics rays) in the problem. A different version of the theory dealt with a degenerate plasma case, when two eigenvalues of the dielectric

tensor vanished simultaneously in some plasma region. In this case, the wave problem could be again reduced to a set of ODEs along the rays. However, in contrast to a *single* amplitude equation in the nondegenerate case, the degeneracy yielded *two coupled* amplitude equations, corresponding to the wave components along the two zero-eigenvectors.

In addition to these multidimensional theories, there existed an independent line of research of the so called *mode conversion* problem, where a wave propagating in a nonuniform plasma could pass a resonance and excite another wave, so that far away from the localized resonant region, the two waves propagate independently. Two approaches to this problem were adopted in the early 80s, both limited to one-dimensional problems only. The first approach by Stix and Swanson [64] was based on a generalization of the tunneling problem. They associated a *single*, higher order ODE for the wave amplitude to mode conversion events and studied its asymptotic solutions via the Laplace transform. The second approach, suggested by Cairns and Lashmore-Davies [65], related the mode conversion problem to local dispersion relations of the form $[k - k_a(x)][k - k_b(x)] - \eta^2 = 0$. This equation allowed to guess a system of coupled differential equations for the amplitudes of the two modes, i.e.

$$\begin{aligned} dA_a/dx - ik_a(x)A_a &= i\eta A_b, \\ dA_b/dx - ik_b(x)A_b &= i\eta^* A_a, \end{aligned}$$

to describe the mode conversion from, say mode a to mode b at the resonance point x_0, such that $k_a(x_0) = k_b(x_0)$. Both approaches seem to be heuristic and unrelated and did not suggest a mechanism of taking into account higher dimensionality and such geometric optics effect as the ray divergence. Nevertheless, the approach of Cairns and Lashmore-Davies appeared to be more physical, because it could be also associated with the dispersion matrix

$$\mathbf{D} = \begin{pmatrix} k - k_a & \eta \\ \eta & k - k_a \end{pmatrix}, \tag{41}$$

taking into account the multi-component structure of the coupled mode propagation problem. Furthermore, having *two* coupled amplitude equations seemed to agree with my knowledge from the geometric optics theory of having a system of coupled amplitude equations in degenerate plasmas ($k_a(x_0) = k_b(x_0)$ is the point of degeneracy in the case of $\epsilon = 0$). Thus, I have concluded at the time that the next goal of the multidimensional WKB theory should be the development of the self-consistent extension of the perturbation approach to mode conversion in space/time varying media. Motivated by this goal, I developed a "renormalized" geometric optics theory of mode conversion in 1985 (see Ref. [50]). The theory was still one-dimensional, and the dielectric tensor was again decomposed in terms of its eigenvectors and eigenvalues. However, the formalism yielded a reduction of the embedded coupled mode system in plasma regions where locally, as one follows a mode along a ray in a nondegenerate plasma, an additional eigenvalue of the dielectric tensor becomes small, i.e., one enters a region of *near degeneracy*. We have also understood that the use of the reduced 3×3 dielectric tensor in mode conversion problems was too restrictive and started employing an unreduced $N \times N$ dispersion matrix $\mathbf{D}$ in studying propagation of multicomponent waves with possible mode conversion events. One year later, we have also suggested another approach to mode conversion [66], which, instead of the decomposition of the large rank dispersion matrix, used a procedure similar to Gauss elimination and related to congruence transformations (see below). The starting point of this approach was a general, unreduced vector-amplitude equation put in the following (conservative) form (the case of $1D$ in x)

$$i\mathbf{D}\cdot\mathbf{A} = -\partial\mathbf{D}/\partial k\cdot d\mathbf{A}/dx - \frac{1}{2}\frac{d(\partial\mathbf{D}/\partial k)}{dx}\cdot\mathbf{A}. \tag{42}$$

We then showed that one can *reduce the order* of this system by eliminating the amplitude components one by one, so that the form of the equation for the remaining components at each reduction step remains unchanged and only $\mathbf{D}$ changes and its rank decreases. By the end of this reduction procedure, one arrived at either a single equation of the aforementioned form for the last remaining amplitude component in the nondegnerate plasma regions (the final reduced $\mathbf{D}$ being a scalar) or remained with a system of two coupled amplitude equations in nearly degenerate regions, where the reduced $\mathbf{D}$ was a 2×2 matrix.

I have presented these theories at the APS meeting in 1985 and, for the first time, met Allan Kaufman, who came to my poster to discuss the new developments in the geometric optics theory. At that time, Allan and his student McDonald were working on the Weyl representation for electromagnetic waves [26] and was interested in our new eikonal-type theories. Shortly after this encounter, in 1986, he invited me to spend my first (after the tenure at the Hebrew University) sabbatical at LBL and join forces in studying the mode conversion paradigm. This sabbatical turned out to be most successful and resulted in three important extensions of the theory. Firstly, we have developed a self-consistent *multidimensional* mode conversion theory [67], yielding the generic form of coupled mode equations embedded in multicomponent wave propagation problems:

$$\begin{aligned} V_a\cdot\partial A_a/\partial x - i(x\cdot R^a)A_a &= i\eta A_b,\\ V_b\cdot\partial A_b/\partial x - i(x\cdot R^b)A_b &= i\eta^* A_a, \end{aligned} \tag{43}$$

where $x = (t, \overrightarrow{r})$ was defined around the resonance, $V_{a,b}$ and $R^{a,b}$ were the 4-group velocities and space/time dispersion coefficients associated with the two modes at the resonance, while η was the off-diagonal element of the associated 2×2 dispersion matrix. We have also found the asymptotic solution of this system predicting a simple formula for the transmission coefficient of the incident wave flux through the mode conversion region, $T = \exp(-2\pi\,|\eta|^2\,/\,|B|)$, with B being the Poisson bracket $B = R^a\cdot V_b - R^b\cdot V_a$. Secondly, we have put the above-mentioned *order reduction* procedure [66] on a firm mathematical bases in 4D space/time varying plasmas showing its relation to congruence transformations [51]. In the same work we have also proved the generic nature of mode conversion events in multicomponent multidimensional wave propagation problems. Finally, in the same year, seeking mathematical simplicity and compactness, Allan developed a phase-space theory of multidimensional mode conversion [68].

On my return to Jerusalem after the sabbatical, I started working on the effect of the nonlinear frequency/wave vector shifts in mode conversion and three wave interactions, leading to a new *autoresonant* wave interaction paradigm [69]. At the same time, Allan and his colleagues (R. Littlejohn, E. Tracy, A. Brizard, J. Morehead, G. Flynn, and A. Jaun), continued working on further developing the linear mode conversion theory and its applications. Some of their important results are the wave-kinetic formulation of incoherent mode conversion [70], the normal forms for linear mode conversion [43], the oceanographic mode conversion [37], the inroduction of the ray helicity concept [88, 71], and applications to tokamak plasmas [17].

> In summary, the above mentioned developments during a relatively short period of time (1985-1987), have changed and shaped the research interests of Allan Kaufman, myself and other people for the next two decades. I feel very fortunate of being a part of these developments."

This ends Lazar's summary of how he contributed to Allan's conversion to mode conversion.

6. Phase space methods in mode conversion

Allan's subsequent work on mode conversion came to rely heavily upon ray phase space ideas. He was led to this formulation upon noting that the transmission coefficient involved the Poisson bracket of the two dispersion functions for the resonant waves. This suggested that symplectic geometry was playing a role that needed to be understood. In addition, he drew heavily upon the (then) recent work of Littlejohn on wavepacket dynamics and the metaplectic transformation.

The phase space viewpoint was first presented in a paper with Friedland [68] which we will discuss in some detail momentarily because the ideas introduced have informed all of the authors' subsequent work with Allan in this area. But, first, let's consider why mode conversion is naturally viewed in ray phase space. A more complete discussion of Allan's thinking on these matters can be found in [19]. This paper is not well known, but shows how far along Allan's thoughts were at that time.

Consider a wave packet moving in more than one spatial dimension in a nonuniform plasma. The envelope is transported by the local group velocity, while the phase evolves according to the local phase velocity. The directions associated with these local velocities are, in general, completely unrelated. Mode conversion is a resonance between two modes that local have the same *phase velocity*, but the group velocities are not required to have any particular relation to one another. Furthermore, in general the coupling coefficient will depend upon the location in phase space, which leads to the possibility of *ray helicity* and conical intersections of dispersion manifolds, rather than 'avoided crossings'. These conical intersections are generic in higher dimension and cannot occur in one dimensional problems. This will also be discussed in the following sections. We first discuss briefly Littlejohn's wavepacket formalism and the metaplectic transformation.

6.1. Littlejohn's wave packet theory and the metaplectic group

Here we discuss related work by Allan's former student Robert Littlejohn because the ideas will become of central importance in Allan's later work on mode conversion. This was independent work of great beauty and is based upon Weyl's insight that all operators, including evolution operators associated with wave equations, can be decomposed into phase space shifts using irreducible representations of the Heisenberg-Weyl group (this is an example of what is now called *non-commutative harmonic analysis*). Each phase space shift has a simple action and if the wave function is decomposed on a wave packet basis, the full wave evolution now looks like that of a gas of particles [72].

In addition to shifts on phase space, Littlejohn considers linear canonical transformations. These arise in the study of wave packet evolution because when the hamitonian flow is linearized about a reference orbit, the nearby orbits evolve in time via a family of linear canonical transformations. Linear canonical transformations are of the form (x and k should be thought of as n-dimensional column vectors, and A, B, C and D are $n \times n$ real matrices):

$$\begin{pmatrix} q \\ p \end{pmatrix} = \begin{pmatrix} \mathsf{A} & \mathsf{B} \\ \mathsf{C} & \mathsf{D} \end{pmatrix} \begin{pmatrix} x \\ k \end{pmatrix} \tag{44}$$

Requiring this to be a canonical transformation implies the the matrix is symplectic. The symplectic matrices form a group. The Heisenberg-Weyl group and the linear symplectic group are related to one another in a manner analogous to translations and rotations in n-dimensional configuration spaces.

Consider now wavefunctions on configuration space $\psi(x)$. Translations in x are easy to understand. Translations in k are multiplications by $\exp(ik \cdot x)$. Therefore, the Heisenberg-Weyl group is represented by general combinations of shifts in x, multiplication by a plane wave, and multiplication by an overall constant phase. How is the group of linear canonical transformations to be represented in this space? By analogy with rotations, there is a set of unitary operators acting on the Hilbert space of wavefunctions ψ which satisfy the same product law as the symplectic matrices. This set of operators forms a group, and the transformations are called the *metaplectic transformations*. They should be thought of as generalizations of Fourier transforms. They are extremely useful because they allow us to take any pair of operators that are linear in $\hat{x}$ and $\hat{k}$ that do not commute and convert them into a 'position' and 'momentum' pair. In the appropriate representation, this can then be used to greatly simplify the local wave equation, as will be discussed in the following sections.

6.2. Plasma echoes and the Budden problem

In a series of papers with his graduate student, Huanchun Ye, Allan pursued the use of these new ideas in the context of gyroresonance absorption. As part of Huanchun's PhD dissertation, Lagrangian Lie transform methods were used to introduce a gyrokinetic description of the plasma and to isolate those regions of the system where gyroresonance was occurring. Following a suggestion of Friedland's, the crossing of a gyroresonance was treated as a mode conversion between a collective plasma wave (such as a magnetosonic wave) and a 'gyroballistic' wave (a solution of the undriven gyrokinetic equation). The entire process can then be viewed in ray phase space. The gyroballistic waves have the interesting property that they propagate slowly in x-space (they are tied to the drift orbits), but can have their primary propagation in k-space. This can be understood by considering a family of gyrating ions in a nonuniform magnetic field. A phase pattern imprinted on these local 'clocks' will evolve because of the nonuniformity, which leads to the evolution in k-space. These ideas lead to an appealing explanation of the phenomenon of linear ion-cyclotron echos, which had been observed by spacecraft [73, 74].

The idea that a Budden-type resonance could be treated in phase space as a two-step conversion process emerged from this work and has been used by our group in a variety of settings ever since.

6.3. Incoherent mode conversion and the use of metaplectic methods

One of the authors (Tracy) recalls first becoming aware of Allan Kaufman's work while a graduate student at the University of Maryland in the early 1980's. Word of Allan's work came through Steve McDonald when he arrived as a post-doc to work with Ed Ott, and through a seminar given by Celso Grebogi. While my thesis work concerned inverse scattering theory and nonlinear waves, I found the ideas in Steve and Celso's work very interesting and wanted to learn more. Upon graduation, I applied to work with Allan but he already had a post-doc (Shayne Johnston). I decided to visit Berkeley anyway during a visit to the West Coast in the spring of 1984 as it would give me a chance to meet him. As it turned out, Allan was unable to be there when I visited, but I remember the kindness of Wulf Kunkel and Robert Littlejohn, who showed me around campus and arranged for my visit in Allan's absence. I wound up accepting a position at William and Mary, but kept abreast of what Allan was up to and finally got to meet him at the APS-DPP and Sherwood meetings. In the late 80's, I was collaborating with Allen Boozer, then at William and Mary, on some problems in geometric optics and decided that I wanted to take the opportunity to make contact again with Allan Kaufman. So, at the APS meeting in 1989 I suggested to Allan that I come for a visit to Berkeley so we could work on problems of joint interest. Allan was kind enough to accept. I visited first in the winter of 1990 and we have been collaborators ever since.

The first problem we worked on concerned multidimensional mode conversion. We chose this because Allan was still pondering the implications of his work with Lazar. Although we have strayed into other areas since then, resonance crossing in one variation or another has been the major theme of our collaboration.

The first task was to compute the converted wave amplitude using Allan and Lazar's normal form (the Physics Letter they had recently published only treated the transmission coefficient). As Lazar and Allan did in their Physics Letter, we used metaplectic ideas to simplify the coupled PDEs. We verified that the resulting S-matrix was unitary. Having the full S-matrix available meant that it was possible to treat the situation with incoming disturbances on both channels, hence the outgoing disturbances would be an interference pattern. Because Steve and Allan's work on wave kinetic equations for incoherent waves was still on our minds it was natural for us to ask whether we could include conversion effects in the wave-kinetic equations. We accomplished this phenomenologically by insisting that away from conversion the two wave types satisfy the action transport equation (the wave kinetic equation), but when one of the rays of type a punctures the dispersion manifold of waves of type b the action has to be split between the two rays [70]. This approach implicitly assumes incoherence because the interference pattern is ignored using a handwaving argument of the type that Marshall Rosebluth used to call a 'random phrase approximation'. Action conservation is satisfied, however.

In order to put these results on a somewhat firmer footing, we then revisited the solutions of the local 2×2 wave equation and, without making an assumption of incoherence, asked what the full solutions looked like in detail. Instead of taking the solutions in the q- or p-representations (which were simple) and using an inverse metaplectic transformation to bring them back to x-space (which looked hard in a technical sense, and it was, see [75] for the details), we used the fact that the Wigner function was invariant under metaplectic transformations. This allowed us to compute the Wigner function in whatever representation was easiest. The result showed that, near conversions, the Wigner function has a very complicated interference structure on the ray phase space. Because the Wigner function involves integrals over the whole space-time, parts of the Wigner function are non-causal. However, when an integration is carried out across the dispersion manifold, one recovers causal behavior for the action density [18].

6.4. The Berkeley plasma theory group

While Allan and Gene were beginning their collaboration, the second author (Alain Brizard) had arrived at Berkeley and joined the very active ferment of ideas being discussed. Alain's work in gyrokinetics and action principles is discussed in a separate article, and his work on mode conversion will be covered in a moment.

At this time (in the early 90's), Allan, Alain and Robert Littlejohn shared a set of offices in Building 4 at LBNL. Robert and Allan's students were also in Building 4, so there was a very definite buzz of activity. Some of the work Robert was pursuing at this time was directly relevant to what we were working on, so we always took advantage of this situation to ask Robert and his students for comments and suggestions. During these years, it was a great pleasure to get to know some very outstanding young scientists: Steven Creagh, Chris Jarzynski, Greg Flynn, Kevin Mitchell, Huanchun Ye, Dan Cook, Jim Morehead, Stefan Weigert, Matt Cargo and Matthias Reinsch.

Robert's work with Greg Flynn was of particular interest to us because it concerned the basic theory of mode conversion [43], with applications not only in plasma [76], but also in quantum systems [77]. One particularly impressive result concerned the appearance of conical intersections, rather than avoided crossings, in multidimensional mode conversion. These cannot occur in one dimensional systems and are a truly new feature of the higher-dimensional problem.

Journal of Physics: Conference Series **169** (2009) 012008 doi:10.1088/1742-6596/169/1/012008

They occur because the off-diagonal term of the local 2×2 wave equation depends upon the point in phase space about which the expansion is taken. Therefore, generically, there will be points in the phase space where the off-diagonal coupling is zero. At such conversions, the entire dispersion matrix vanishes at linear order and the local geometry of the dispersion manifolds in phase space has a cone structure. Exploiting the fact that the dispersion matrix is a 2×2 hermitian matrix, Robert and Greg used a decomposition in terms of the Pauli matrices:

$$\mathsf{D} = \begin{pmatrix} D_{11} & D_{12} \\ D_{12}^{*} & D_{22} \end{pmatrix} = \sum_{\mu} B_{\mu} \sigma^{\mu}. \tag{45}$$

with $B_0 = (D_{11} + D_{22})/2$, $B_1 = \mathrm{Re}(D_{12})$, $B_2 = \mathrm{Im}(D_{12})$, and $B_3 = (D_{11} - D_{22})/2$. With this reparametrization of the dispersion matrix, the determinant takes the form:

$$\det(\mathsf{D}) = B_0^2 - B_1^2 - B_1^2 - B_1^2. \tag{46}$$

The condition $\det(D) = 0$ is now seen to imply that the four-vector (B_0, B_1, B_2, B_3) must lie on the 'light cone'. The components B are functions on phase space, and the apex of the light cone is the point where all entries of D vanish. (Strictly speaking, it is only a point in four-dimensional phase spaces. In higher dimensions the local geometry will locally be the product of a cone and a plane.) This is important because, near a conical intersection, the outgoing wave field will have a contribution (or deletion, depending upon the channel) that has the form of a gaussian beam rather than being eikonal.

Littlejohn and Flynn then developed some very beautiful arguments concerning the universality of the S-matrix connection coefficients, showing that they are almost completely determined by symmetry requirements [34].

Through the 90's and into the 00's, we continued to extend this line of effort in various ways, as described in the next section.

A tutorial article in Physics of Plasmas appeared in 2003 that summarized our understanding of ray-based methods in mode conversion up to that point [78].

7. Extensions, elaborations and related work by other groups

7.1. Minority gyroresonance and the computation of the reflected wave in tokamak geometry

Lazar Friedland had pointed out that when a collective wave crosses a minority gyroresonance the mathematical form of the equations wave look very much like that for a mode conversion. This is often overlooked because the particle disturbance, as described by the linearized evolution equation for the perturbed particle distribution, is usually eliminated in favor of the electric field. This leads to the usual singular denominator associated with wave-particle resonance. If one does not eliminate the particle distribution, but instead treats the electric field and particle distribution on an equal mathematical footing, then the resulting equations are similar to the 2×2 form of a mode conversion. The reflected wave is emitted when the disturbance in the minority-ion distribution has evolved until it satisfies the collective wave dispersion relation once more (this is the same idea that appears in Huanchun Ye's work with Allan). Strictly speaking, this is only true when the minority ions are cold. When they are warm, then there is an infinite family of minority 'modes' that take part in the conversion. Each is resonant at a different spatial position because of Doppler effects and the spatial variation in the local magnetic field strength.

Allan's student, Dan Cook, pursued these issues and reported on them in a series of papers. The work proceeded along two fronts: a full kinetic treatment of conversion to gyroballistic waves in one-dimension, and conversion to cold gyroballistic waves, but now in two spatial dimensions.

In [79], the conversion to a continuum of gyroballistic waves was considered for a one-dimensional slab model. Tor Flå was visiting the Berkeley group at the time and joined in this work. This paper gave the first analytic formulas for energy transfer from a magnetosonic wave to the minority Bernstein wave and the absorption profile, for plasma heating via minority gyroresonance. The absorption profile stems from: (1) gyroresonant damping of the Bernstein wave; (2) phase mixing of the kinetic residue. As part of this work, Dan did a direct comparison between the semi-analytic theory and a direct numerical simulation, getting good agreement in the limit of small minority fraction. Computation of the absorption profile using these ideas required developing a means to extract the minority-ion Bernstein wave which arises due to the self-consistent fields associated with the gyroballistic waves. Strictly speaking, when these interactions are included one is no longer dealing with gyroballistic wave, but with Case-van Kampen modes. The method of Case and van Kampen was first developed in a uniform plasma. The extension to non-uniform plasma was due to Bateman and Kruskal, which could be adapted to the present setting.

In [80] the reflected wave in tokamak geometry was constructed using a very simple conceptual model. Phase-space methods were used for the conversion of an incident magnetosonic wave field to a continuum of gyroballistic waves, for their propagation along guiding-center orbits in the poloidal plane of a tokamak, and for their conversion to the two-dimensional magnetosonic reflection field. The wave fronts and amplitude of that field were obtained explicitly. This work did not include a full kinetic treatment.

The further development of these ideas to compute the conversion to the minority-ion Bernstein wave in tokamak geometry required extending the Bateman-Kruskal theory to multidimensions [81]. The generalization to many dimensions and non-trivial geometries required several important new developments: In tokamak geometry particles can be trapped, an effect that is absent in the slab model. Also, the ray propagation dynamics for both the free gyroballistic waves and the collective minority-ion Bernstein wave is far more complicated than in the slab model. In particular, a resonance zone was identified wherein the gyroballistic waves interact strongly and cannot be treated as free. We used the Weyl calculus to construct a local form of the self-consistent gyroballistic equation within the resonance zone. This reduced equation was simplified via a metaplectic transformation. After this simplification, the equation was shown to be of Case-van Kampen type with weak non-uniformities; hence there are no true Case-van Kampen eigenfunctions. Using the Bateman-Kruskal approach, a local Case-van Kampen basis was constructed and the initial-value problem solved.

7.2. Multiple conversion

Because the conversion is local in phase space, this means that multiple conversions can be dealt with in a modular fashion if they are well separated. It is important to realize that this separation is required to be in *phase space*, not in configuration space. It is possible for two conversions to lie over one another in x-space yet be well separated in phase space because they occur for very different k's. This is, for example, the case with a Budden-type resonance, as discussed in [82, 83]. In these Budden-type of double-conversions, the signal pathways are all open, hence there is no possibility of interference effects occurring due to rays following multiple pathways.

In some cases, however, multiple conversion can occur in such a way that signal can follow closed circuits in phase space. An example occurs in one spatial dimension when the ray paths are curved and cross (like intersecting parabolas). If the two crossing are sufficiently far apart, they can be treated in modular fashion again and the global input/output relationships can be deduced by repeated application of the single-step conversion formulas [84]. These effects can be particularly dramatic if the two intersection waves are of opposite energy, in which case there is amplification at each crossing and the double-crossing becomes an absolute instability. A simple model with multiple conversions was considered in [85], where it was shown that to get the proper phase matching between the conversions one must go beyond WKB order and include the leading order effects of the coupling. Once this is done, the interference effects are correctly captured using the modular form. More recently, we have considered the possibility of recirculation that can occur if a tertiary wave is present, in addition to the primary and secondary waves usually associated with a Budden resonance. This tertiary wave might be due, for example, to the presence of energetic particles [86].

Multiple conversion has been an ongoing interest of the group, and we have returned to it several times. For example, when considering the effects of flow on the magnetosonic to ion-hybrid conversion, we found that a virtual cavity can be created and multiple conversions can occur if the flow has sufficient shear [87]. Because of the local Doppler shift caused by the flow, the dispersion relations of waves are modified, and a single Budden resonance may be replaced by a triple resonance. The occurrence of such triplication depends on the layer's parameters: its location, maximum flow speed, and radial width. We used a standard one-dimensional slab model, with cold-plasma response in the ion-gyrofrequency range, with the example of conversion of a magnetosonic wave to ion-hybrid waves. When triplication occurs, wave energy is converted at three locations instead of one. This possibility must be taken into account in designing conversion scenarios for heating, for current-drive, or for flow-drive.

7.3. *Wave emission in nonuniform plasmas*

Consider the emission of waves by sources, such as resonant particles. This is usually dealt with by assuming the plasma is uniform near the localized source. From this, one computes the emission spectrum using Fourier methods and propagates this disturbance to the far field then switches to WKB. But, what if the nonuniformity of the medium cannot be neglected near the source? Allan noted that in phase space this is essentially a form of resonance crossing and in a series of papers we used the tools we had developed for analyzing mode conversion in phase space to attack this problem. In [54] we considered the case of emission of a scalar wave by a gyrating particle, while in [55, 56] we considered the case of emission that occurs within a mode conversion region.

7.4. *Ray helicity*

Robert and Greg's conical intersection work [77], and their work on the universality of the connection coefficients [34], inspired us to pursue more fundamental issues on conversion. In particular, we considered how to simplify the local 2×2 wave equation as much as possible by various changes of representation, guided by the proposition that for a ray-based quantity to have a fundamental physical meaning, it must be invariant under the following two groups of transformations, which are used to construct solutions: congruence transformations (which involve linear combinations of components of the multicomponent wave field) and canonical transformations (which act on the ray phase space). In [88] we showed that for conversion between two waves there is a new invariant not previously discussed: the intrinsic helicity of the ray. If the determinant of the dispersion matrix is used to generate rays, then it is possible to show that the ray motion in phase spaces of dimension higher than two will always be a combination of hyperbolic and elliptic motions, and that the combined motion is locally a helix, hence a one-dimensional model is not adequate. This ray helicity can be invariantly defined and enters into the S-matrix coefficients (unpublished).

In [89] we applied these ideas to the conversion from a magnetosonic wave to an ion-hybrid wave in tokamak geometry. A cold-plasma model was introduced in this paper which exhibits ray helicity in conversion regions where the density and magnetic field gradients are significantly nonparallel. For illustration, such regions are identified in a model of the poloidal plane of a DT tokamak plasma. In each conversion region, characterized by a six-sector topology, rays in the sector for incident and reflected magnetosonic waves exhibit significant helicity. A detailed analytic and numerical study of helical rays in this sector was then developed for a "symmetric-wedge" model.

7.5. Applications beyond plasma physics

Darryl Holm suggested that we should consider applications of mode conversion ideas in the theory of equatorial waves and led us to some earlier work on upwelling in the Gulf of Guinea off the coast of West Africa. A resonance had been identified between an equatorially trapped Rossby-Kelvin wave (also known as a Yanai wave) and a coastally trapped Kelvin wave. The coupling of the waves is due to the overlap of their eigenfunctions (normal modes in y, the meridional coordinate). After deriving coupled mode equations from a variational principle, which gives the coupling constant between the two modes at the conversion point, we then used the S-matrix formulas obtaining an analytic expression for the wave-energy conversion coefficient, in terms of the wave frequency and the scale length of the thermocline depth [37]. These ideas were later picked up and used by Tailleux and McWilliams in several papers [38, 39].

Cally and co-workers have used phase space methods in their study of magnetohelioseismology [40] where ray splitting has been found in certain models of the solar atmosphere.

7.6. Algorithms for numerical computation

In addition to studies of more foundational issues, our group has also pursued the development of a practical numerical algorithm [90, 17, 75]. A ray tracing algorithm must allow for the ray splitting that occurs at a mode conversion, and requires solutions of the following subproblems: (1) how to detect conversion regions while following rays; (2) once conversion is detected, how to fit to a generic saddle structure in ray phase space associated with the most common type of conversion (the 'avoided crossing' type); (3) given the saddle structure, how to carry out a local projection of the full vector wave equation onto a local two-component normal form that governs the two resonantly interacting waves. This determines both the uncoupled dispersion functions and the coupling constant, which in turn determine the uncoupled WKB solutions; (4) given the normal form of the local two-component wave equation, how to find the particular solution that matches the amplitude, phase, and polarization of the incoming ray, to the amplitude, phase, and polarization of the two outgoing rays: the transmitted and converted rays. This work was carried out in collaboration with André Jaun, who had previously worked on the development of full-wave codes to study conversion in tokamak plasmas, and more recently with Steve Richardson.

7.7. Theoretical foundations and work in progress

The phase space approach that our group has pursued is based upon the Weyl symbol calculus. Our colleague Nahum Zobin has developed a new version of this theory that is based upon group theoretic ideas. This leads to a generalization of the Weyl symbol, and the usual definition (24) is then seen to be a special case. This new formulation of the symbol calculus is based upon the representation theory of the Heisenberg-Weyl group. The theory then leads very naturally to the *phase space path integral* (a variant of the Feynman path integral, which is on configuration space). The path integral arises when we ask the following question: suppose we wish to solve an evolution equation of the Schrödinger form:

$$i\hbar\partial_t|\psi\rangle = \hat{H}|\psi\rangle. \tag{47}$$

Suppose, in addition, we know the symbol of the operator $\hat{H}$, denoted $H(x,k)$. The standard way to solve (47) is to exponentiate $\hat{H}$ to find the evolution operator $\hat{U}_t = \exp(i\hat{H}t)$. But, this requires a full knowledge of the eigenvalues and eigenvectors of $\hat{H}$ (so that we can diagonalize it). This information is usually not available. The only other general method for computing the

Journal of Physics: Conference Series **169** (2009) 012008 doi:10.1088/1742-6596/169/1/012008

evolution operator is to use the symbol calculus. In this approach, we first construct the symbol of $\hat{U}_t$ using our knowledge of the symbol $H(x, k)$. Knowledge of the symbol is tantamount to knowledge of the operator. When this calculation is carried out, the path integral emerges. In practice the path integral cannot be computed because it involves summations over all possible paths, the vast majority of which are wild. However, one usually invokes the limit $\hbar \to 0$ and considers the leading term in the asymptotics of the path integral, which usually are dominated by the classical (ray, or WKB) paths. See Steve Richardson's dissertation for details [27].

Suppose we apply these ideas to 2×2 wave equations. In that case $\hat{H}$ is a 2×2 operator-valued matrix and $|\psi\rangle$ a two-component spinor. The symbol of $\hat{H}$ is a 2×2 hermitian matrix. The 'classical limit' of the path integral in this circumstance is very subtle when there is a resonance. In fact, the path integral suggests that it is the *diagonals* of the symbol of $\hat{H}$, that is $H_{11}(x, k)$ and $H_{22}(x, k)$ that are acting as the ray Hamiltonians, not the determinant or the eigenvalues. The choice of representation becomes very important, because in some representations the asymptotic limits will be cleaner than others. This led us to a new definition of the *normal form* for 2×2 wave equations: it is a representation where the symbols on the diagonal Poisson-commute with the symbols on the off-diagonals. In this representation, the diagonals can be interpreted as the uncoupled ray Hamiltonians and the off-diagonals are constant along rays generated by the diagonals (because of the condition of Poisson commutativity). This makes it natural to interpret the off-diagonals as coupling constants. We are currently pursuing these ideas, and attempting to construct practical algorithms for generating this normal form representation while following a ray.

Our group is currently working with Allan on several problems: the direct comparison of ray-based and full-wave calculations for multidimensional conversion, the development of phase space representations for full-wave calculations, and the application of modular methods to compute full-wave maps in tokamak geometry. Our extended group of collaborators has been joined by two Ph.D. students, Yanli Xiao and David Johnston, who are working on various aspects of these problems.

8. Summary

It is clear from this selective survey that Allan's years of work in plasma wave theory have been highly creative and productive. Many new ideas were brought into the field either directly by Allan, or through his close association with Robert Littlejohn. Their students have become productive researchers as well. For the two authors of the current paper, the plasma theory group at Berkeley has always been a stimulating place to visit and we have very much enjoyed being a part of the group.

Acknowledgments

This work, which took place over many years, has been supported by a variety of sources. Allan Kaufman's work was supported by the US-DOE Office of Fusion Energy Sciences and the Office of Basic Energy Sciences. The authors wish to also acknowledge support from the NSF-DOE Partnership in Basic Plasma Physics as well as support from the US-DOE Office of Fusion Energy Sciences.

We would also like to acknowledge another important contributor to this effort: Allan's wife Louise. Throughout the last twenty years, she has been a constant source of support and encouragement, and we thank her for her warm hospitality. During our visits she has always been a gracious hostess. And, most important, she tolerated our tendency to talk shop at dinner with good humor and made us feel at home.

References

[1] B. I. Cohen, A. N. Kaufman and K. M. Watson, "Beat heating of a plasma," PRL **28** (1972) 581.
[2] A. J. Brizard, "Local Manley-Rowe relations for noneikonal wave fields," PRL **74** (1995) 4567.
[3] A. N. Kaufman, "Quasilinear diffusion of an axisymmetric toroidal plasma," Physics of Fluids **15** (1972) 1063.
[4] H. Ye and A. N. Kaufman, "Self-consistent theory for ion gyroresonance," Physics of Fluids, **4**B (1992), 1735-53.
[5] G. R. Smith and A. N. Kaufman, "Stochastic acceleration by a single particle in a magnetic field," PRL **34** (1975) 1613.
[6] J. R. Cary and A. N. Kaufman, "Ponderomotive effects in collisionless plasma: A Lie transform approach," Physics of Fluids **24** (1981) 1238.
[7] S. Johnston, A. N. Kaufman and G. L. Johnston, "Beat Hamiltonians and generalized ponderomotive forces in hot magnetized plasma," J. Plasma Physics **20** (1978) 365.
[8] A. N. Kaufman, "Dissipative Hamiltonian systems: a unifying principle," Physics Letters **100**A (1984) 419.
[9] P. J. Morrison, "A paradigm for joined Hamiltonian and dissipative systems," Physica **18**D (1986) 410-19.
[10] Grmela, M. "Bracket formulation of dissipative time evolution equations," Physics Letters **111**A (1985) 36-40.
[11] C. Grebogi, A. N. Kaufman and R. G. Littlejohn, "Hamitonian theory of ponderomotive effects of an electromagnetic wave in a nonuniform magnetic field," PRL **43** (1979) 1668.
[12] P. L. Similon, A. N. Kaufman and D. D. Holm, "Ponderomotive Hamiltonian and Lyapunov stability for magnetically confined plasma in the presence of RF field," Phys. Lett. **106**A (1984) 29.
[13] P. L. Similon, A. N. Kaufman and D. D. Holm, "Oscillation center theory and ponderomotive stabilization oflow-frequency plasma modes," Physics of Fluids **29** (1986) 1908.
[14] A. N. Kaufman and D. D. Holm, "The Lie-transformed action principle: relativistically covariant wave propagation and self-consistent ponderomotive effects," Phys. Lett. **105**A (1984) 277.
[15] G. B. Whitham, *Linear and nonlinear waves* (Wiley, New York, 1974).
[16] B. Boghosian, PhD. Dissertation, University of California at Berkeley, 1986.
[17] A. Jaun, E. R. Tracy and A. N. Kaufman, Plasma Physics and Controlled Fusion **49** (2007) 43.
[18] E. R. Tracy, A. N. Kaufman, "Metaplectic formulation of linear mode conversion," Phys. Rev. E 48 (1993) 2196.
[19] A.N. Kaufman, "Phase-space action principles, linear mode conversion, and the generalized Fourier transformation", in *Nonlinear and chaotic phenomena in plasmas, fluids and solids*, W. Rozmus and J. A. Tuszynski, eds. (World Scientific, New Jersey, 1991).
[20] R. G. Littlejohn and J. M. Robbins, "New way to compute Maslov indices", Phys. Rev. A **36**, 2953 (1987).
[21] E. E. Nikitin and J. Troe, "70 years of Landau-Teller theory for collisional energy transfer. Semiclassical three-dimensional generalizations of the classical collinear model," Physical Chemistry Chemical Physics **10** (2008) 1483-1501.
[22] E. E. Nikitin, "Nonadiabatic transitions: what we learned from the old masters and how much we owe them," Annual Reviews of Physical Chemistry **50** (1999) 1-21.
[23] R. G. Littlejohn, "Conical intersection, light cones and mode conversion," to appear in the Proceedings of the Edward Teller Centennial Symposium, May 2008, Livermore, CA.
[24] H. Weyl, *The theory of groups and quantum mechanics* (Dover, New York, 1950).
[25] S. W. McDonald, "Phase space representations of wave equations with applications to the eikonal approximation for short-wavelength waves", Phys. Reports **158**, 337 (1988).
[26] S.W. McDonald and A.N. Kaufman, "Weyl representation for electromagnetic waves: the wave kinetic equation," Phys. Rev. A **32**, 1708 (1985).
[27] A. S. Richardson, "Group theoretical foundations of path integrals with application in mode conversion," (PhD. dissertation, William and Mary, 2007).
[28] R. L. Littlejohn, "The semiclassical evolution of wave packets", Physics Reports **138** (1986) 193-291.
[29] S. W. McDonald and A. N. Kaufman, "Spectrum and eigenfunctions for a Hamiltonian with stochastic trajectories," PRL **42** (1979) 1189.
[30] S. W. McDonald and A. N. Kaufman, "Wave chaos in the stadium: statistical properties of short-wave solutions of the Helmholtz equation," Physical Review A **37** (1988) 3067.
[31] S. W. McDonald and A. N. Kaufman, "Weyl representation for electromagnetic waves: the wave kinetic equation," Physical Review A **32** (1985) 1708-13.
[32] A. N. Kaufman, "Phase-space Lagrangian action principle and the generalized $K-\chi$ theorem," Physical Review A **36** (1987) 982.
[33] A. N. Kaufman, H. Ye and Y. Hui, "Variational formulation of covariant eikonal theory for vector waves," Physics Letters **120**A (1987) 327.

[34] R. G. Littlejohn and W. G. Flynn, "Geometric phases and the Bohr-Sommerfeld quantization of multicomponent wave fields", PRL **66** (1991) 2839.
[35] See, for example, *RF Power in Plasmas*, T. K. Mau and J. deGrassie, eds., AIP Conference Proceedings (American Institute of Physics, Melville, NY, 2001) Vol. **595**.
[36] K. G. Budden, *The propagation of radio waves* (Cambridge University Press, New York, 1985).
[37] A. N. Kaufman, J. J. Morehead, A. J. Brizard and E. R. Tracy, "Mode conversion in the Gulf of Guinea", J. Fl. Mech. **394** (1999) 175.
[38] R. Tailleux and J. C. McWilliams, J. C., "Linear resonance, WKB breakdown, and the coupling of Rossby waves over slowly varying topography", in IUTAM symposium on "Advances in Mathematical Modelling of Atmosphere and Ocean Dynamics" (Kluwer Academic Publishers, 2001.) pp. 259-264.
[39] R. Tailleux and J. C. McWilliams, "Energy propagation of long, extratropical Rossby waves over slowly varying zonal topography", Journal of Fluid Mechanics 473, (2002) 295-319.
[40] P. S. Cally, "Dispersion relations, rays and ray splitting in magnetohelioseismology", Phil. Trans. Roy. Soc. A **364** (2006) 333.
[41] E. E. Nikitin and S. Ya. Umanskii, *Theory of slow atomic collisons* (Springer Series in Chemical Physics, Vol. 30) (Springer-Verlag, New York, 1984).
[42] G. G. Raffelt, *Stars as laboratories for fundamental physics*, (University of Chicago, Chicago, 1996), Chapt. 8.
[43] W. G. Flynn and R. G. Littlejohn, "Normal forms for linear mode conversion and Landau-Zener transitions in one dimension", Annals of Physics **234**, 334 (1994).
[44] A. N. Kaufman, E. R. Tracy, J. J. Morehead and A. J. Brizard, "The dissipative Budden problem: effect of converted-wave damping on primary-wave reflection", Phys. Lett. A **252**, 43 (1999).
[45] A. K. Ram, A. Bers, S. D. Schultz and V. Fuchs, "Mode conversion of fast Alfvén waves at the ion-ion hybrid resonance", Phys. Plasmas **3**, 1976 (1996).
[46] D. G. Swanson, *Theory of mode conversion and tunneling* (Wiley, New York, 1998).
[47] V. Fuchs, K. Ko and A. Bers, "Theory of mode conversion in weakly inhomogeneous plasma", *Phys. Fluids* **24**, 1251 (1981).
[48] R. A. Cairns, "The role of negative energy waves in some instabilities of parallel flows", J. Fluid Mech. **92** (1979) 1.
[49] I. B. Bernstein and L. Friedland, "Geometric optics in space and time varying plasmas", in *Handbook of Plasma Physics*, edited by M. N. Rosenbluth and R. Z. Sagdeev (North-Holland, Amsterdam, 1984), Vol. 1.
[50] L. Friedland, "Renormalized geometric optics description of mode conversion in weakly inhomogeneous plasma", Phys. Fluids **28**, 3260 (1985).
[51] L. Friedland and A. N. Kaufman, "Congruent reduction in geometric optics and mode conversion," Phys. Fl. **30** (1987) 3050.
[52] R. G. Littlejohn, "Landau-Zener transitions in two dimensions", in *Path Integrals from meV to MeV*, Proceedings of the 4th International Conference, Tutzing, Bavaria, May 18-21 (1992), edited by Hermann Grabert, *et al.* (World Scientific, Singapore, 1993).
[53] E. R. Tracy and A. N. Kaufman, "Metaplectic formulation of linear mode conversion", Phys. Rev **48**E (1993) 2196.
[54] E. R. Tacy, A. N. Kaufman and Y.-M. Liang, "Wave emission by resonance crossing", Phys. Plasmas **2** (1995) 4413.
[55] Yu. Krasniak and E. R. Tracy, "Emission from mode conversion regions in multi-dimensions: a new diagnostic probe for non-uniform media", Phys. Lett. A **248** (1998) 235.
[56] Yu. Krasniak and E. R. Tracy, "Emission from mode conversion regions", in *Radio Frequency Power in Plasmas, 13th Topical Conference* (AIP Conference Proceedings, No. 485, 1999).
[57] N. Nassiri-Mofkham and B. Sabzevari, "Mode conversion in plasmas with two-dimensional inhomogeneities," J. Plasma Physics **72** (2005) 71.
[58] P. J. Braam and J. J. Duistermaat, "Normal forms of real symmetric systems with multiplicity", Indag. Mathem., N.S. Vol. 4(4), 407 (1993).
[59] P. J. Braam and J. J. Duistermaat, "Normal forms of real symmetric systems with multiplicity", in *Panoramas of Mathematics*, Banach Center Publications, Vol. 34, Institute of Mathematics, Polish Academy of Sciences, Warszawa (1995).
[60] Y. Colin de Verdière, "The level crossing problem in semi-classical analysis I: the symmetric case", Annales de l'Institute Fourier, Special Issue: Proceedings of Fréderic Pham's Congress, **53** (2003) 1023-1054.
[61] Y. Colin de Verdière, "The level crossing problem in semi-classical analysis I: the symmetric case", Annales de l'Institute Fourier, Special Issue: In Honor of Louis Boutet de Monvel, **54** (2004) 1423-1441.
[62] C. Emmrich and A. Weinstein, "Geometry of the transport equation in multicomponent WKB

approximations," Commun. Math. Phys. **176** (1996) 701.

[63] C. Emmrich and H. Römer, "Multicomponent Wentzel-Kramers-Brillouin aproximation on arbitrary symplectic manifolds: a star product approach," J. Math. Phys. **39** (1996) 3530.

[64] T. Stix and D.G. Swanson, in Ref. [49], Chap. 2.4.

[65] R.A. Cairns and C.N. Lashmore-Davies, "A unified theory of a class of mode conversion problems" Phys. Fluids **26**, 1268 (1983).

[66] L. Friedland and G. Goldner, "Reduction of order in the geometric optics of plasmas" Phys. Fluids **29**, 4073 (1986).

[67] L. Friedland, G. Goldner, and A.N. Kaufman, "Four-dimensional eikonal theory of linear mode conversion" Phys. Rev. Lett. **58**, 1392 (1987).

[68] A.N. Kaufman and L. Friedland, "Phase space solution of the linear mode-conversion problem," Phys. Lett. A **123**, 387 (1987).

[69] L. Friedland, 'Spatial autoresonance: enhancement of mode conversion due to nonlinear phase locking'," Phys. Fluids B **4**, 3199 (1992); "Autoresonant three-wave interactions," Phys. Rev. Lett. **69**, 1749 (1992); "Autoresonant excitation and evolution of nonlinear waves: the variational approach," Phys. Rev. E **55** 1929 (1997).

[70] E.R. Tracy and A.N. Kaufman, "Wave-kinetic formulation of incoherent mode conversion," Phys. Rev. Lett. **64**, 1621 (1990).

[71] A. N. Kaufman, E. R. Tracy and A. J. Brizard, "Helical rays in two-dimensional resonant wave conversion," Phys. Plasmas **12** (2005) 022101.

[72] R. G. Littlejohn, "The semiclassical evolution of wavepackets," Physics Reports **138** (1986) 195.

[73] H. Ye and A. N. Kaufman, "Wave propagation in k-space and the linear ion-cyclotron echo," Physical Review Letters **60** (1988) 1642.

[74] H. Ye and A. N. Kaufman, "Analytic solution for second-harmonic gyroresonant absorption and mode conversion ," Physical Review Letters **61** (1988) 2762.

[75] E. R. Tracy, A. N. Kaufman and A. N. Jaun, "Local fields for asymptotic matching in multdimensional mode conversion," Physics of Plasmas **14** (2007) 82102/1-28.

[76] W. G. Flynn and R. G. Littlejohn, "RF heating of the ionosphere: and example of generic linear mode conversion," *AIP Conference Proceedings* **289** (1994) 261-4.

[77] R. G. Littlejohn and W. G. Flynn, "Phase integral theory, coupled wave equations, and mode conversion", Chaos **2**, 149 (1992).

[78] "Ray-based methods in multidimensional linear wave conversion," E. R. Tracy, A. N. Kaufman and A. J. Brizard, Phys. Plasmas 10 (2003) 2147.

[79] D. R. Cook, A. N. Kaufman, E. R. Tracy and T. Fla, "Collective wave spin-off and the gyroballistic continuum in gyronresonant absorption," Physics Letters **175**A (1993) 326.

[80] D. R. Cook, A. N. Kaufman, A. J. Brizard, H. Ye and E. R. Tracy, "Two-dimensional reflection of magnetosonic radiation by gyroballistic waves: an analytical theory," Physics Letters **178**A (1993) 413.

[81] E. R. Tracy, A. J. Brizard and A. N. Kaufman, "Generalized Case-van Kampen modes in a mutldimensional non-uniform plasma with application to gyroresonance heating," J. Plasma Physics **55** (1996) 449-86.

[82] A. N. Kaufman, J. J. Morehead, A. J. Brizard and E. R. Tracy, "Modular approach for conversion to the ion-hybrid wave and α-gyroresonance," in *Radio Frequency Power in Plasmas*, 12th *Topical Conference*, AIP Conference Proceedings **403** (1997) 359.

[83] A. N. Kaufman, E. R. Tracy, J. J. Morehead and A. J. Brizard, "The dissipative Budden problem: effect of converted wave damping on primary-wave reflection," Physics Letters **252** A (1999) 43.

[84] A. J. Brizard, J. J. Morehead, A. N. Kaufman and E. R. Tracy, "Double-crossing mode conversion in a nonuniform plasma," Physics of Plasma **5** (1998) 45-59.

[85] Y.-M. Liang, J. J. Morehead, D. R. Cook, T. Fla and A. N. Kaufman, "Multiple mode conversion: an analytical and numerical comparison," Physics Letters **193** A (1994) 82-88.

[86] A. J. Brizard, A. N. Kaufman and E. R. Tracy, "Recirculation in multiple wave conversion," Physics of Plasmas **15** (2008) 082116.

[87] A. N. Kaufman, A. J. Brizard and E. R. Tracy, "Triplicate Budden resonance in the presence of sheared flow," AIP Conference Proceedings **787** (2005) 146.

[88] E. R. Tracy and A. N. Kaufman, "Ray Helicity: geometric invariant for multidimensional resonant wave conversion," Phys. Rev. Lett. **91** (2003) 130402.

[89] A. N. Kaufman, E. R. Tracy and A. J. Brizard, "Helical rays in two-dimensional resonant wave conversion," Physics of Plasmas **12** (2005) 22101-1-10.

[90] E. R. Tracy, A. N. Kaufman, A. Jaun, "A ray-based algorithm for multi-dimensional linear conversion," Phys. Lett. A290 (2001) 309.

Plasma Theory, Wave Kinetics, Nonlinear Dynamics: KaufmanFest 2007 IOP Publishing
Journal of Physics: Conference Series **169** (2009) 012009 doi:10.1088/1742-6596/169/1/012009

Analyzing Charged Particle Beams using Lie Transform Perturbation Theory

Kiran G. Sonnad
Center for Beam Physics, Lawrence Berkeley Lab., Berkeley California USA
E-mail: kgsonnad@gmail.com

John R. Cary
John R. Cary, Center for Integrated Plasma Studies and Dept. of Physics, University of Colorado, Boulder. Colorado USA

Abstract. This paper briefly reviews the use of Lie transform perturbation theory applied to the dynamics of charged particle beams. Several details not provided here can be found in [1] and [2]. The work analyzes influences due to nonlinear components in external focusing of beams, a regime where the Courant-Snyder invariants are broken [3] for single particle dynamics. The Lie transform analysis is also used to analyze beams with space charge effects, which satisfy the Vlasov -Poisson system of equations. Equilibrium distributions may be obtained even in the nonlinear focusing regime, where simple solution such as the Kapchinskij-Vladimirskij [4] are no longer valid. The work utilizes results that were obtained earlier as part of the research activities of Prof. Allan Kaufman and members of his group.

1. Lie Transform Perturbation Methods for Hamiltonian Systems - A Brief Review

In this section, we outline the Hamiltonian perturbation method described in detail in Ref. [5] and more briefly in Ref. [6]. This method is based on previous work [7, 8, 9, 10, 11] that introduced Lie transform theory as a convenient method to perform Hamilton perturbation analysis. The Lie transformation is defined with respect to a phase space function w such that it satisfies the following Poisson bracket relationship,

$$\frac{d\mathbf{Z}}{d\epsilon} = \{\mathbf{Z},\ w(\mathbf{Z}(\mathbf{z}, t, \epsilon), t, \epsilon)\}, \tag{1}$$

where $\mathbf{Z} = (\mathbf{P}, \mathbf{Q})$ is a phase space vector representing the generalized positions and momenta of the system, w is the Lie generating function and ϵ is a continuously varying parameter such that $\mathbf{Z}(\epsilon = 0) = \mathbf{z}$, the original phase space vector. The above relationship resembles Hamilton's equation with respect to a "Hamiltonian", w and "time", ϵ. This guarantees that the transformation is canonical for all values of ϵ.

The Lie operator L is defined such that it performs a Poisson bracket operation with respect to w. Symbolically,

$$L = \{w,\ \}. \tag{2}$$

A transformation operator T is defined such that its role is to replace the variables of a function by the new canonical variables. For the identity function this is simply,

$$T\mathbf{z} = \mathbf{Z}(\mathbf{z}, \epsilon, t). \tag{3}$$

The operator T is analogous to the "evolution" operator with respect to ϵ. Using Eq. (1) it can be verified that T satisfies

$$\frac{dT}{d\epsilon} = -TL. \tag{4}$$

For a similar relationship involving the inverse transformation operator T^{-1}, we differentiate the equation $TT^{-1} = 1$ and use the above equation to obtain

$$\frac{dT^{-1}}{d\epsilon} = T^{-1}L. \tag{5}$$

The transformed Hamiltonian K can be expressed in terms of the original Hamiltonian $\mathcal{H}$ as

$$K(\epsilon) = T^{-1}(\epsilon)(\mathcal{H}) + T^{-1}(\epsilon)\int_0^{\epsilon} d\epsilon' T(\epsilon')\frac{\partial w}{\partial t}(\epsilon'). \tag{6}$$

This expression was obtained by Dewar [10].

Using Eq. (4), and the Deprit [9] series expression for the operators L and T, gives us the following recursion relationship

$$T_n = -\frac{1}{n}\sum_{m=0}^{n-1} T_m L_{n-m}. \tag{7}$$

Similarly, using Eq. (5), and the Deprit series expression for the operators L and T^{-1}, the inverse transformation operator may be expressed recursively as

$$T_n^{-1} = \frac{1}{n}\sum_{m=0}^{n-1} L_{n-m}T_m^{-1} \tag{8}$$

It may be noted that when L, T and T^{-1} act upon any phase space function, they are expressed in the form of Poisson brackets, which are independent of the canonical variables used. This makes the whole formulation canonically invariant.

To obtain the nth order perturbation equation, we premultiply Eq. (6) by T and differentiate with respect to ϵ to obtain

$$\frac{\partial T}{\partial \epsilon}K + T\frac{\partial K}{\partial \epsilon} = \frac{\partial H}{\partial \epsilon} + T\frac{\partial w}{\partial t} \tag{9}$$

Using Eq. (4) to eliminate $\partial T/\partial\epsilon$ (with $dT/d\epsilon \to \partial T/\partial \epsilon$, since here T also depends explicitly on t) and premultiplying by T^{-1},

$$\frac{\partial w}{\partial t} = \frac{\partial K}{\partial \epsilon} - LK - T^{-1}\frac{\partial H}{\partial \epsilon} \tag{10}$$

Inserting the series expansions and equating like powers of ϵ, we obtain in nth order,

$$\frac{\partial w_n}{\partial t} = nK_n - \sum_{m=0}^{n-1} L_{n-m}K_m - \sum_{m=1}^{n} mT_{n-m}^{-1}H_m. \tag{11}$$

To third order, this equation yields,

$$K_0 = H_0, \tag{12}$$

$$\frac{\partial w_1}{\partial t} + \{w_1, H_0\} = K_1 - H_1, \tag{13}$$

$$\frac{\partial w_2}{\partial t} + \{w_2, H_0\} = 2(K_2 - H_2) - L_1(K_1 + H_1), \tag{14}$$

$$\begin{aligned}\frac{\partial w_3}{\partial t} &+ \{w_3, H_0\} = 3(K_3 - H_3) - L_1(K_2 + 2H_2) \\ &-L_2(K_1 + \frac{1}{2}H_1) - \frac{1}{2}L_1^2 H_1.\end{aligned} \tag{15}$$

The expression $\frac{\partial w_n}{\partial t} + \{w_n, H_0\}$ is the variation of w_n along the unperturbed trajectory described by H_0. In the following section we use this perturbation scheme to perform time averaging. To do this, we set $H_0 = 0$ which reduces the variation of w_n along a trajectory to a partial derivative with respect to t. Thus, instead of integrating along the unperturbed trajectory, we simply perform an integration over time to determine w_n. At each order, K_n is chosen such that it cancels the terms that average to a nonzero value over fast oscillations. As a result, the corresponding value of w_n will have a zero average. This is necessary to prevent w_n from being secular (unbounded) in time [5]. For a systematic derivation of all these relationships one may refer to Ref. [5] where they are given up to fourth order in ϵ.

2. Application to a Linear Sinusoidal Focusing System

As an illustration and a test for the validity of the method, we perform the analysis for a linear periodic focusing system. The same example was used in Ref. [12] for the method developed in that paper. The single particle Hamiltonian associated with such a system is given by

$$H = \frac{p^2}{2} + \frac{kq^2}{2}\sin(\omega t). \tag{16}$$

This Hamiltonian also describes the motion of a particle in systems such as the Paul trap and the ponderomotive potential. We apply Eqs. (12 - 15) to perform the averaging. As explained in the previous section, we set $H_0 = 0$ and $H_1 = H$. From Eq (12) we get,

$$K_0 = H_0 = 0, \tag{17}$$

Applying the first order relationship, Eq (13), we get

$$\frac{\partial w_1}{\partial t} = K_1 - \frac{p^2}{2} - \frac{kq^2}{2}\sin(\omega t). \tag{18}$$

The third term on the right averages to zero with respect to time. In order that the net result average to zero, we require

$$K_1 = \frac{p^2}{2}. \tag{19}$$

Since w_1 is relevant only up to an additive constant, it is sufficient to evaluate the indefinite integral to determine w_1, hence

$$w_1 = \frac{kq^2}{2\omega}\cos(\omega t). \tag{20}$$

The second order equation Eq.(14) gives

$$\frac{\partial w_2}{\partial t} = 2K_2 - \frac{2kpq}{\omega}\cos(\omega t). \tag{21}$$

Journal of Physics: Conference Series **169** (2009) 012009 doi:10.1088/1742-6596/169/1/012009

Since the second term on the right side averages to zero, we choose

$$K_2 = 0, \tag{22}$$

and so,

$$w_2 = -\frac{2kqp}{\omega^2}\sin(\omega t). \tag{23}$$

Applying the third order relationship, Eq. (15) then gives

$$\frac{\partial w_3}{\partial t} = 3K_3 + \frac{3p^2k}{\omega^2}\sin(\omega t)$$
$$-\frac{k^2q^2}{\omega^2}\sin^2(\omega t) - \frac{k^2q^2}{2\omega^2}\cos^2(\omega t). \tag{24}$$

Note that the third and fourth terms on the right side do not average to zero. In order that they cancel, we set

$$K_3 = \frac{1}{4}\frac{k^2q^2}{\omega^2} \tag{25}$$

and as a result,

$$w_3 = -\frac{3p^2k}{\omega^3}\cos(\omega t). \tag{26}$$

Collecting the nonzero terms, the transformed Hamiltonian is now given as a function of the new variables by

$$K = \frac{P^2}{2} + \frac{\Omega^2Q^2}{2} \tag{27}$$

where $\Omega = k/\sqrt{2}\omega$. This is the Hamiltonian for a harmonic oscillator with solution

$$Q(t) = Q(0)\cos(\Omega t) + \frac{P(0)}{\Omega}\sin(\Omega t), \tag{28}$$

$$P(t) = P(0)\cos(\Omega t) - \Omega Q(0)\sin(\Omega t). \tag{29}$$

To transform back to the original coordinate system, we use the operator T^{-1} for which we need to know L up to the desired order. The operators L_n can be expressed in terms of the values of w_n as

$$L_1 = \{\frac{kQ^2}{2\omega}\cos(\omega t),\ \}, \tag{30}$$

$$L_2 = \{-\frac{2kQP}{\omega^2}\sin(\omega t),\ \}, \tag{31}$$

$$L_3 = \{-\frac{3kP^2}{\omega^3}\cos(\omega t),\ \}. \tag{32}$$

Using these to perform the inverse transformation as described we get, up to third order,

$$q = Q + \frac{kQ}{\omega^2}\sin(\omega t) + \frac{2kP}{\omega^3}\cos(\omega t), \tag{33}$$

$$p = P + \frac{kQ}{\omega}\cos(\omega t) - \frac{kP}{\omega^2}\sin(\omega t)$$
$$+\frac{1}{3}\frac{k^2}{\omega^3}Q\sin(\omega t)\cos(\omega t). \tag{34}$$

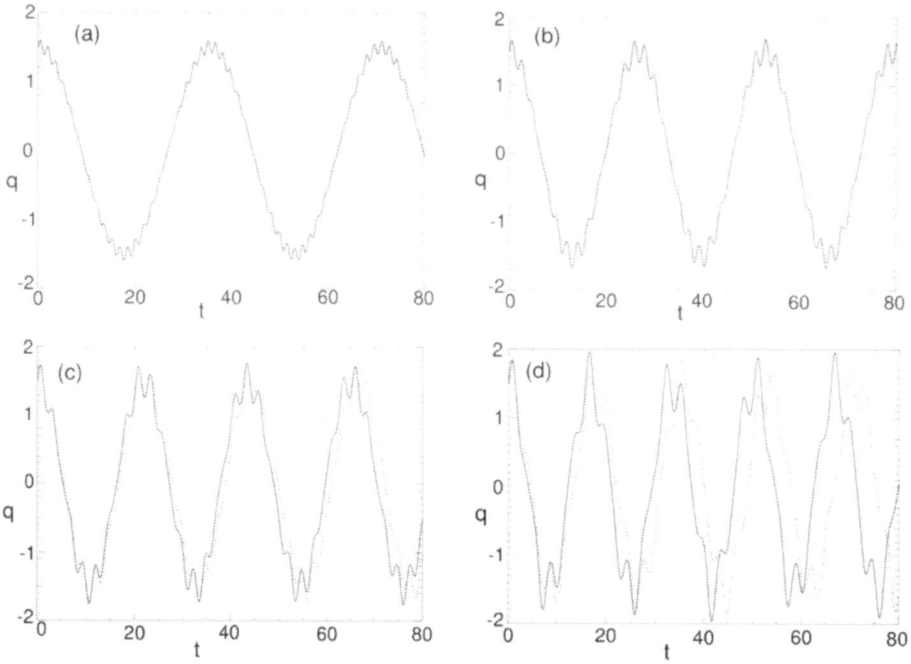

Figure 1. q vs t with $k = 1$, $\omega =$ (a) 4, (b) 3, (c) 2.5 and (d) 2. The solid line represents the numerical solution

The above solution is compared with calculations from a fourth order symplectic integrator [13, 14] and is shown in Fig. (1). The parameters used were the same as those used in Ref. [12]. The accuracy of the approximate solution compares well with that obtained by Channell [12] using a different method. That is, the solution given by Eqs. (33) and (34) overlaps well with the numerical solution for $k/\omega^2 = 1/16$ and the accuracy gradually decreases with decreasing ω.

3. Application to Single Particle Dynamics in Accelerators

For a single particle moving in a particle accelerator under the influence of quadrupoles and sextupoles, the Hamiltonian can be expressed in cylindrical coordinates as

$$\begin{aligned} H &= \frac{1}{2}(p_r^2 + \frac{l^2}{r^2}) + \frac{1}{2}\kappa_2(s) r^2 \cos(2\theta) \\ & \frac{1}{3}\kappa_3(s) r^3 \cos(3\theta + \alpha). \end{aligned} \tag{35}$$

The variable s is the distance along the axis, which is equivalent to time for constant axial velocity. The momentum in the radial direction is p_r and l is the angular momentum. The values of $\kappa_2(s)$ and $\kappa_3(s)$ depend upon the strength of the quadrupole and sextupole magnets respectively and also the velocity of the particle in the axial direction. The angle α depends upon the relative values of a_2 and b_2 which is determined by the orientation of the sextupoles with respect to the quadrupoles. We use normalized units in which the charge and mass of the particle are unity. It is assumed that the Hamiltonian is periodic in s with periodicity S, *ie*, $\kappa_2(s+S) = \kappa_2(s)$ and $\kappa_3(s+S) = \kappa_3(s)$. It is further assumed that the average of $\kappa_2(s)$ and $\kappa_3(s)$ over a period S is zero. That is,

$$\langle \kappa_2 \rangle = \frac{1}{S}\int_s^{s+S} \kappa_2(s) ds = 0 \tag{36}$$

and the same for κ_3. The angle brackets $\langle\cdots\rangle$ denote an average over one period in the rest of this paper.

Employing the same procedure as before, it can be shown [1] that the transformed Hamiltonian gives

$$K = \frac{1}{2}(P_R^2 + \frac{l^2}{R^2})\frac{1}{2}\langle(\kappa_2^{\rm I})^2\rangle R^2$$
$$-\frac{1}{3}\langle\kappa_2^{\rm I}\kappa_3^{\rm I}\rangle R^3\cos(\Theta+\alpha) + \frac{1}{2}\langle(\kappa_3^{\rm I})^2\rangle R^4. \tag{37}$$

In order for K to be independent of Θ, the transformed azimuthal variable, $\langle\kappa_2^{\rm I}\kappa_3^{\rm I}\rangle$ must vanish. The simplest way to achieve this with real quadrupoles and sextupoles is as shown in Fig 2. One could extend the analysis to more complex systems as well. The separation between the quadrupoles and sextupoles is represented by a term,

$$\psi = \frac{2\pi\Delta s}{S} \tag{38}$$

where Δs is the spatial distance between the two of them. The averaged Hamiltonian K is independent of Θ when $\psi = 90^o$, *ie*, when the sextupoles are placed halfway between two quadrupoles of opposite sign.

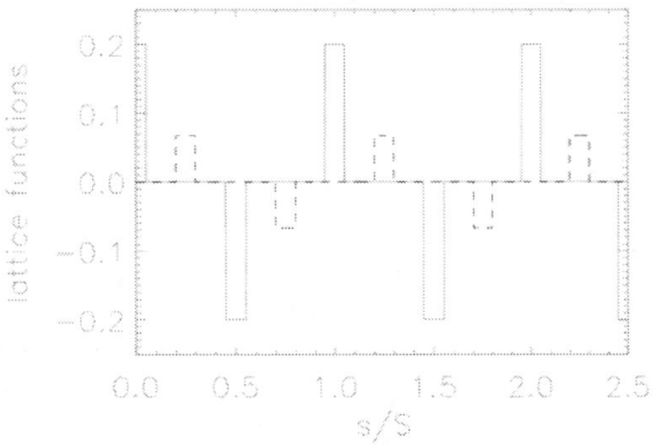

Figure 2. A step function lattice that will lead to a near integrable condition. The shorter steps represent the sextupole function $\kappa_3(s)$ while the higher ones the quadrupole function $\kappa_2(s)$.

The dynamic aperture is defined as the volume in phase space in which all particles remain confined throughout their trajectories in the accelerator. The calculations in this section estimate the projection of the dynamic aperture onto various phase space planes for different values of ψ.

In Fig. (3), the left side represents the initial phase space positions of confined particles and the ones on the right represent the unconfined particles from the same initial distribution. It is important to plot the confined and unconfined particles separately in order to ensure that there is no overlap between the two regions, which is true in these simulations. This is expected because all the phase space variables other than those shown in the respective plot were set to zero. Given that the dynamic aperture allows only confined particles and not a mixture of the two, the left side plots represent the projection of the dynamic aperture onto the respective plane.

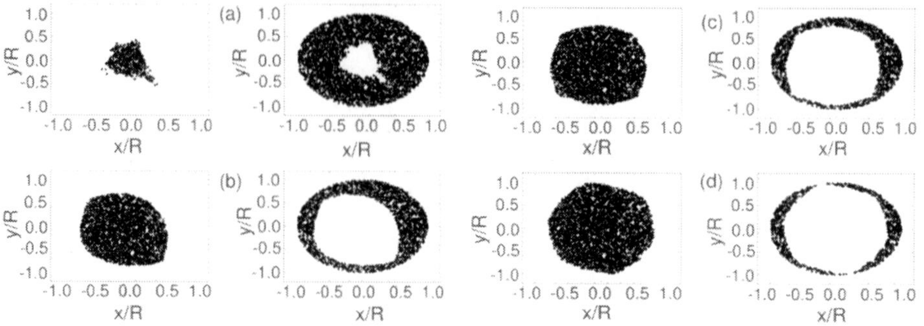

Figure 3. Initial distribution of confined and unconfined particles lying on the x-y plane for $\psi =$ (a) 0°, (b) 30° (c) 60°, (d) 90°

4. Generalization to Beams with Space Charge Effects

In the study of self consistent systems such as beams with space charge effects, it is often required to solve for the the phase space density $f(\mathbf{q},\mathbf{p})$ which satisfies the Vlasov equation,

$$\frac{df}{ds} = \frac{\partial f}{\partial t} + \{H, f\} = 0. \tag{39}$$

Lie transform methods have been used to analyze systems governed by the Vlasov equation in the past (for example see Ref[15]. In self consistent systems, since the Hamiltonian includes the space charge potential, H itself will be a function of f. In many cases, it is possible to decompose H as

$$H = H^e + H^{sc}, \tag{40}$$

where H^e is the term arising from external forces and H^{sc} arises from space charge forces. In Vlasov-Poisson systems, H^{sc} is proportional to the electrostatic potential ϕ which satisfies Poisson's equation,

$$\nabla^2\phi(\mathbf{q}, s) = -\epsilon^m \frac{q}{\epsilon_0} \int d\mathbf{p} f(\mathbf{q}, \mathbf{p}, s) \tag{41}$$

where q is the charge of the particle and ϵ_0 is the free-space permittivity. In Eq. (41), we have put in the ordering parameter, ϵ^m for maximal ordering. Since we are solving for beam equilibria, this factor will be selected so that the lowest-order potential is of the same order as the lowest-order averaged confining forces.

If we define a function

$$F(\mathbf{Q}, \mathbf{P}, s) = f(\mathbf{q}(\mathbf{Q}, \mathbf{P}, s), \mathbf{p}(\mathbf{Q}, \mathbf{P}, s), s), \tag{42}$$

then it may be easily verified that

$$f(\mathbf{q}, \mathbf{p}, s) = TF(\mathbf{q}, \mathbf{p}, \mathbf{s}). \tag{43}$$

From the property of canonical invariance of the Vlasov equation (see for example Ref [12]), we then have

$$\frac{dF}{dt} = \frac{\partial F}{\partial s} + \{K, F\} = 0. \tag{44}$$

Using the perturbation expansion terms of the transformation operator T we can in principle expand f in terms of F using Eq. (43) in order to obtain the phase space distribution in the original coordinate system. This gives,

$$f = f_0 + \epsilon f_1 + \epsilon^2 f_2 + \epsilon^3 f_3 + \ldots \tag{45}$$

where

$$f_n(\mathbf{q}, \mathbf{p}, \mathbf{s}) = T_n F(\mathbf{q}, \mathbf{p}, \mathbf{s}) \tag{46}$$

Combining Eqs. (45) and (41), ϕ may be expressed as a decomposition of its different order terms such that,

$$\phi = \epsilon^m(\phi_m + \epsilon\phi_{m+1} + \epsilon^2\phi_{m+2} + \epsilon^3\phi_{m+3} + \ldots) \tag{47}$$

where

$$\nabla^2\phi_n(\mathbf{q}, s) = -\frac{q}{\epsilon_0}\int d\mathbf{p} f_{n-m}(\mathbf{q}, \mathbf{p}, s) \tag{48}$$

The different terms in the electrostatic potentials represent contributions arising from the different orders in the perturbation of the charge density. Thus, the lowest nonzero term in ϕ corresponds to the lowest nonzero term in f.

The motivation to carry out the procedure outlined in this section is based on the presumption that it is possible to perform a canonical transformation from $(\mathbf{q}, \mathbf{p}, s)$ to $(\mathbf{Q}, \mathbf{P}, s)$ such that the resulting phase space function, $F(\mathbf{q}, \mathbf{p}, s)$ is easier to solve for than $f(\mathbf{q}, \mathbf{p}, s)$. It is then more convenient to first solve for F and then transform back to f rather than solving directly for f. In this study, we solve for an equilibrium, where F is independent of s. This is achieved through a transformation that depends on s. If the transformation is perturbative, which is true in this context, f may be obtained only up to a desired order. It has been shown in the appendix that the terms higher than the lowest nonzero term in f do not change the total number of particles in the distribution. Thus, the total charge is conserved even upon truncation of the series expansion in f.

The Hamiltonian for a beam with space charge effects is given by

$$\begin{aligned} H &= \frac{1}{2}(p_r^2 + \frac{l^2}{r^2}) + \frac{1}{2}\kappa_2(s)r^2\cos(2\theta) \\ &\quad \frac{1}{3}\kappa_3(s)r^3\cos(3\theta + \alpha) + \frac{q}{m_0 v_s^2\gamma^3}\phi(r, \theta, s). \end{aligned} \tag{49}$$

Over here, m_0 is the rest mass, q the charge and v_s the axial velocity of the particle, and γ is the Lorentz relativistic factor. Using the same averaging procedure described in this paper, and requiring that the system be azimuthally symmetric, one could obtain the transformed Hamiltonian, which is given by

$$\begin{aligned} K &= \frac{1}{2}(P_R^2 + \frac{\mathcal{L}^2}{R^2}) + \frac{1}{2}\langle(\kappa_2^{\rm I})^2\rangle R^2 + \frac{1}{2}\langle(\kappa_3^{\rm I})^2\rangle R^4 \\ &\quad -\frac{1}{3}\langle\kappa_2^{\rm I}\kappa_3^{\rm I}\rangle R^3\cos(\Theta + \alpha) + \frac{q}{m v_s^2\gamma^3}\phi_3 \end{aligned} \tag{50}$$

As before, the superscript I represents an integration with respect to "time", s with the same conditions as described in the previous section.

The averaged Hamiltonian now has an external potential independent of s allowing one to compute an equilibrium distribution, which is a steady state solution of Eq. (44) given by

$$F = F(K(\mathbf{Q}, \mathbf{P})). \tag{51}$$

Transforming F to f up to the desired order based on the procedure described in the previous section would then yield a near equilibrium distribution function in the original coordinate system.

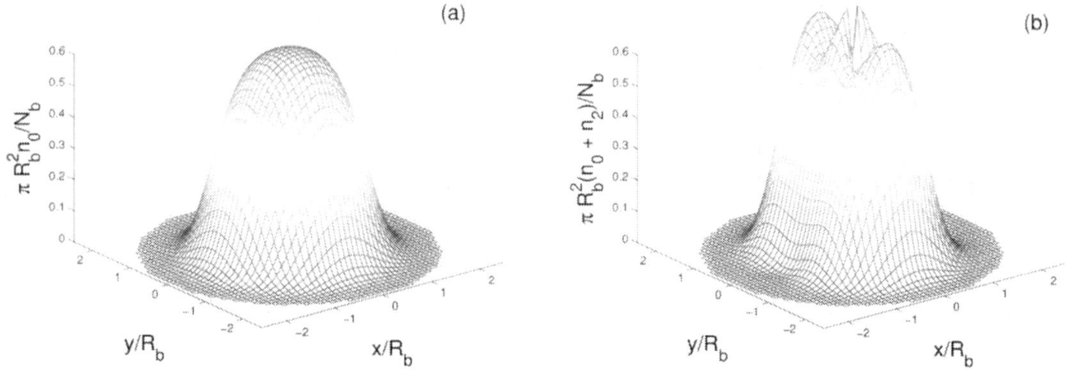

Figure 4. Purely linear focusing (a) plot number density of n_0, (b)$(n_0 + n_2$ located in the middle of an x-focusing quadrupole

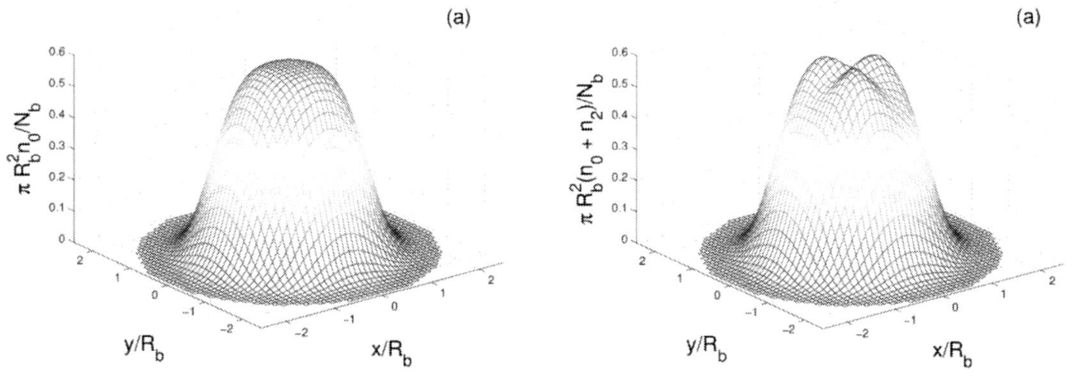

Figure 5. Nonlinear focusing, (a) plot of number density n_0, and (b) $(n_0 + n_2)$ located in the middle of a sextupole

A thermal equilibrium distribution is then given by

$$F(\mathbf{P}, \mathbf{Q}) = \frac{\hat{n}}{2\pi \mathcal{T}_0} \exp[-K/\mathcal{T}_0] = \frac{\hat{n}}{2\pi \mathcal{T}_0} \exp\{ -[\frac{1}{2}(P_R^2 + \frac{\mathcal{L}^2}{R^2}) + V(R) + \frac{q}{mv_s^2\gamma^3}\phi_0]/\mathcal{T}_0\} \quad (52)$$

where $\hat{n}$ is the density at $R = 0$, $\mathcal{T}_0$ represents the temperature of the beam, in units of the Hamiltonian and the external potential term $V(R)$ is given by

$$V(R) = \frac{1}{2}\langle(\kappa_2^{\mathrm{I}})^2\rangle R^2 + \frac{1}{2}\langle(\kappa_3^{\mathrm{I}})^2\rangle R^4 \quad (53)$$

Transferring this distribution to the original phase space variables, one could get the "s" dependence of the distribution upto the desired order.

Figure 4 shows how the "s" dependence for a purely linear external focusing case changes when one takes into account perturbations over the averaged equilibrium distribution. The

Journal of Physics: Conference Series **169** (2009) 012009 doi:10.1088/1742-6596/169/1/012009

averaged distribution in position space, which is Fig 4(a) is symmetric about θ, while Fig 4(b) shows the dependence of the distribution with respect to θ inside a quadrupole magnetic field.

Figure 5 shows how the "s" dependence for a nonlinear linear external focusing case changes when one takes into account perturbations over the averaged equilibrium distribution. The averaged distribution in position space (5)(a) is symmetric about θ, while (5)(b) shows the dependence of the distribution with respect to θ inside a sextupole magnetic field.

It is clear that the symmetry of the distribution reflects the symmetry of the external focusing element. A detailed formulation showing the derivation of the distribution will be presented in a paper under preparation [2].

5. Summary

Lie Transform theory has proved to be a powerful tool for analyzing certain aspects of beams physics. Work in this area is being continued by us and we hope that the effort will lead to a more optimized designs and better understanding of particle accelerators in the future. Nonlinear components such as sextupoles and octupoles are regularly used in storage rings and optimizing the dynamic aperture can greatly help improve the performance of such facilties. Beams with space charge effects are important in heavy ion fusion experiments and spallation neutron sources. The analysis presented in this paper may even find applications in systems other than particle accelerators.

Refrences

[1] K. G. Sonnad and J. R. Cary, 2004, Phys. Rev E, 69, p 056501
[2] K. G. Sonnad and J. R. Cary, *Near equilibrium beam distribution functions for linear and nonlinear periodic focusing systems*, in preparation.
[3] E. D. Courant and H. S. Snyder, 1958, Annals of Physics, vol 3, p 1
[4] I. Kapchinskij and V. Vladimirskij, 1959, "*Proceedings of the International Conference on High Energy Accelerators and Instrumentation*" (CERN Scientific Information Services, Geneva 1959)", p 274
[5] J. R. Cary, 1981, Physics Reports, vol 79, p 129
[6] A.J. Lichtenberg and M.A. Lieberman, 1992, *Regular and chaotic dynamics* (Springer-Verlag)
[7] G. Hori, 1966, Astron. Soc. Japan, 18, p 287
[8] L. M. Garrido, 1968, J. Math. Phys., 10 p 1045
[9] A. Deprit, 1969, Cel. Mech., 1, p 12
[10] R. L. Dewar, 1976, J. Phys., A9, p 2043
[11] A. J. Dragt and J. M. Finn, 1976, J. Math. Phys., 17, p 2215
[12] P. Channel, 1999, Phys. Plasmas, vol 6, p 982
[13] E. Forest and R.D. Ruth, 1990, Physica D, 43, p 105
[14] J. Candy and W. Rozmus, 1991, J. of Comp. Phys., 92, p 230
[15] J. R. Cary and Allan Kaufman, 1981, Phys. Fluids, 24, p 1238

AUTHOR INDEX